ANNALS OF MATHEMATICS STUDIES

NUMBER 1

ALGEBRAIC THEORY OF NUMBERS

BY

HERMANN WEYL

PRINCETON
PRINCETON UNIVERSITY PRESS
LONDON: HUMPHREY MILFORD
OXFORD UNIVERSITY PRESS
1940

ISBN 0-691-07908-0

Sixth Printing, 1971

Printed in the United States of America
by Princeton University Press, Princeton, New Jersey

PREFACE

These are the authentic notes of a course on Theory of Numbers given in Princeton during the year 1938-1939; I say authentic because they were written down by the lecturer himself. The first two chapters of the course, dealing with the elementary theory of divisibility for ordinary integers and for polynomials of one and several variables, have here been omitted. Where I left off, Professor Chevalley continued with a course on Class Fields. In these notes, some of the material presented by him has been worked into the last chapter, on algebraic numbers, so as to pave the way to the modern theory of class fields and Abelian fields. In Chapter II I have axiomatized Kronecker's approach to the problem of divisibility, which has recently been completely neglected in favor of ideals; the reasons for this procedure are given in the text. The ultimate verdict may be that the one outstanding way for any deeper penetration into the subject is the Kummer-Hensel p-adic theory. In view of the comparative scarcity of books in English on theory of numbers, I hope that this outline of the fundamental arithmetic concepts and facts concerning algebraic fields will be of some use.

Hermann Weyl

The Institute for Advanced Study,
Princeton, New Jersey

A SHORT BIBLIOGRAPHY (BOOKS ONLY)

L. E. Dickson, History of the theory of numbers, Carnegie Institution, 1919-23, 3 vols.

A. A. Albert, Modern Higher algebra, Chicago, 1937

v. d. Waerden, Moderne Algebra, Berlin, 1937 and 1931, 2 vols.

H. Weber, Lehrbuch der Algebra, 2nd vol., Vieweg, 1899

L. E. Dickson, Modern elementary theory of numbers, Chicago, 1939

Hardy and Wright, An introduction to the theory of numbers, Oxford, 1938

Dirichlet-Dedekind, Vorlesungen über Zahlentheorie (4th and and last edition, Braunschweig, 1894!)

Algebraic numbers, Report of the Committee on Algebraic Numbers, Bulletin of the National Research Council, Nos. 28 and 62 (Washington, D.C., 1923 and 1928)

E. Landau, Einführung in die elementare und analytische Theorie der algebraischen Zahlen und der Ideale, Leipzig, 1927

H. Minkowski, Diophantische Approximationen, Leipzig, 1907

E. Hecke, Vorlesungen über die Theorie der algebraischen Zahlen, Leipzig, 1923

D. Hilbert, "Zahlbericht," Gesammelte Abhandlungen, vol. I, Berlin, 1932, No. 7 (pp. 63-539)

Hensel-Landsberg, Theorie der algebraischen Funktionen einer Variabeln, Leipzig, 1902

K. Hensel, Theorie der algebraischen Zahlen, Leipzig, 1908

Krull, Idealtheorie, Ergebn. d. Math. IV 3, 1935

H. Hasse, Bericht über neuere Untersuchungen und Probleme aus der Theorie der algebraischen Zahlkörper, Jahresber. Deutsch. Math. Ver. 35, 1926, 1-55; 36, 1927, 233-311

H. Hasse, Klassenkörpertheorie, Mimeographed Notes, Marburg, 1932-33

CONTENTS

Page

Chapter I. ALGEBRAIC FIELDS 1
1. Finite field. Norm, trace, discriminant . . 1
2. Tower. Analysis of the field equation . . . 6
3. Simple extension 9
4. Relative trace, norm and discriminant . . . 15
5. Removal of the hypothesis of separability . 18
6. The Galois case 21
7. Consecutive extensions replaced by a single one 24
8. Strictly finite fields 28
9. Adjunction of indeterminates 30

Chapter II. THEORY OF DIVISIBILITY (KRONECKER, DEDEKIND) 33
1. Integers 33
2. Our disbelief in ideals 35
3. The axioms 38
4. Consequences 40
5. Integrity in $\kappa(x,y,..)$ over $k(x,y,..)$ . . . 44
6. Kronecker's theory 49
7. The fundamental lemma 54
8. A batch of simple propositions 59
9. Relative Norm of a Divisor 63
10. The Dedekind case 63
11. Kronecker and Dedekind 66

Chapter III. LOCAL PRIMADIC ANALYSIS (KUMMER, HENSEL) 71
1. Quadratic number field 71
2. Kummer's theory: decomposition 75
3. Kummer's theory: discriminant 79
4. Prime cyclotomic fields 80
5. Program 83
6. p-adic and $\mathfrak{p}$-adic numbers 94
7. $\kappa(\mathfrak{p})$ and $\kappa(\mathfrak{P})$ 100
8. Discriminant 107
9. Relative discriminant 111
10. Hilbert's theory of Galois fields. Artin symbol 116

Page

Chapter III (Continued)
11. Cyclotomic field and quadratic law of reciprocity 124
12. General cyclotomic fields 129

Chapter IV. ALGEBRAIC NUMBER FIELDS 141
1. Lattices (old-fashioned) 141
2. Field basis and basis of an ideal 145
3. Norm and number of residues 147
4. Euler's function and Fermat's theorem 150
5. A new viewpoint 153
6. Minkowski's geometric principle 158
7. A fundamental inequality and its consequences: existence of ramification ideals, classes of ideals 163
8. The Dirichlet-Minkowski-Hasse-Chevalley construction of units. 168
9. The structure of the group of units 171
10. Finite Abelian groups and their characters . 175
11. Asymptotic equi-distribution of ideals over their classes 178
12. ζ-function and related Dirichlet series . . . 182
13. Prime numbers in residue classes modulo m . 190
14. ζ-function of quadratic fields, and their application 193
15. Norm residues in quadratic fields 201
16. General norm residue symbol and the theory of class fields 210

Amendments 223

Chapter I

ALGEBRAIC FIELDS

1. Finite Field. Norm, Trace, Discriminant

Let $\varkappa$ be a field and k a subfield of $\varkappa$, so that $\varkappa$ is a field "over k." The elements of $\varkappa$ are denoted by Greek letters and simply called elements, while Roman letters and the word "number" shall for the time being be reserved for the elements of k. Elements may be added and an element multiplied by a number: these operations together with the axioms holding for them constitute $\varkappa$ as a vector space. We assume that this vector space has a finite number n of dimensions; n is then said to be the degree of $\varkappa$ over k,

$$n = [\varkappa:k].$$

We repeat the well-known definition of dimensionality for a vector space, and thus introduce the notion of a (vector) basis. m elements $\lambda_1, \ldots, \lambda_m$ are linearly dependent (with regard to k) if they satisfy a relation

$$a_1\lambda_1 + \ldots + a_m\lambda_m = 0$$

with numbers $a_1, \ldots, a_m$ (in k) which do not all vanish. $\varkappa$ is of degree n if any n + 1 elements of $\varkappa$ are linearly dependent while there exist n linearly independent elements $\omega_1, \ldots, \omega_n$. These form a basis, and every element ξ can be uniquely expressed in the form

$$(1.1) \qquad \xi = x_1\omega_1 + \ldots + x_n\omega_n$$

where the numbers $x_1, \ldots, x_n$ are called the components of ξ. Let $\overset{*}{\omega}_i$ $(i = 1, \ldots, n)$ be any other basis. The $\overset{*}{\omega}_i$ may be expressed in terms of the basis ω_i,

$$(1.2) \qquad \overset{*}{\omega}_i = \sum_k l_{ki}\omega_k,$$

and vice versa. Therefore the matrix $L = \| l_{ik} \|$ must be non-singular. Conversely if $L = \| l_{ik} \|$ is a non-singular matrix, then (1.2) defines a new basis ω_i^* in terms of the original one ω_i. In expressing ξ in terms of the new basis,

$$\xi = x_1^*\omega_1^* + \dots + x_n^*\omega_n^*,$$

we have

$$x_i = \sum_k l_{ik} x_k^*, \tag{1.3}$$

or in the notation of matrix calculus

$$x = Lx^*$$

with x standing for the column of the numbers $x_1, \dots, x_n$.

A linear mapping $\xi \to \eta$ in our vector space carries the elements ω_i of the given basis into elements ω_i' and hence ξ, (1.1), into

$$\eta = x_1\omega_1' + \dots + x_n\omega_n'.$$

If

$$\omega_i' = \sum_k a_{ki}\omega_k$$

one has

$$\eta = y_1\omega_1 + \dots + y_n\omega_n$$

with

$$y_i = \sum_k a_{ik}x_k.$$

Hence the linear mapping is described by a matrix $A = \| a_{ik} \|$ in terms of the given basis. One easily verifies that the same mapping is described by the matrix

$$L^{-1}AL \tag{1.4}$$

in terms of the basis (ω_i^*) which arises from (ω_i) by the transformation L, (1.2).

We now take into account the operation of multiplying any two elements of $\varkappa$. For a given element α the equation

$$\eta = \alpha \cdot \xi$$

defines a linear mapping A: $\xi \to \eta$ in $\varkappa$; A denotes at the same time the matrix expressing this mapping in terms of a given basis (ω_i):

$$(1.5) \qquad \alpha \cdot \omega_i = \sum_k a_{ki}\omega_k, \qquad A = \|a_{ik}\|.$$

The correspondence $\alpha \to A$ is a representation, called the regular representation, i.e., $\alpha \to A$, $\beta \to B$ entails

$$c\alpha \to cA \text{ (c any number)}, \quad \alpha + \beta \to A + B, \quad \alpha\beta \to AB.$$

Moreover the element 1 is represented by the (n-dimensional) unit matrix E. This is a good way of expressing the distributive and associative nature of multiplication. For instance, $\alpha\beta \to AB$ states that the mapping associated with $\alpha\beta$ is obtained by performing the two mappings associated with β and α, one after the other (first β, then α); or

$$(\alpha\beta)\xi = \alpha(\beta\xi).$$

While we started by considering $\varkappa$ as a vector space (first level) it now appears more particularly as an algebra (second level). The peculiarities which characterize a field (third level) among algebras are

(1) the axiom of division: for any $\alpha \neq 0$ there exists an element α^{-1} such that $\alpha \cdot \alpha^{-1} = 1$ (division algebra);

(2) the axiom of commutativity of multiplication.

They are of a considerably more refractory nature than the assumptions characterizing $\varkappa$ as an algebra. For the time being we continue to move on the second level.

Let A be a linear mapping and its matrix, $\|a_{ik}\|$, t an indeterminate. The characteristic polynomial

$$(1.6) \qquad f(t) = t^n - a_1 t^{n-1} + a_2 t^{n-2} - \dots \pm a_n$$

of the mapping A is introduced as the determinant of the matrix tE - A. According to (1.4), f(t) is an invariant of A, namely independent of the basis in terms of which the mapping is expressed as a matrix A. In particular the trace

$$a_1 = \sum_i a_{ii} \text{ and the norm, } a_n = \det(a_{ik})$$

are invariants. The trace of the product of two mappings A, B equals

$$\sum_{i,k} a_{ik} b_{ki}$$

and hence is symmetric in A and B.

We apply these remarks to the linear mapping A: $\xi \to \alpha\xi$, associated by the regular representation with the element α. The trace and norm of A are called trace and norm of α and denoted by

$$S(\alpha), \qquad \mathrm{Nm}\,\alpha$$

respectively. The equations (1.5) or

$$\sum_k (\alpha\delta_{ki} - a_{ki})\omega_k = 0$$

at once show that

$$\det(\alpha\delta_{ki} - a_{ki}) = 0,$$

or that α is a root of the characteristic equation

$$f(t) = \det(tE - A).$$

[$E = \| \delta_{ik} \|$ denotes the unit matrix.] We therefore call f(t) the field equation of α, and we have proved:

Theorem I 1,A. *Every element of $\varkappa$ satisfies a definite algebraic equation of degree n with coefficients in k, its field equation.*

The trace $S(\xi)$ depends linearly on ξ:

$$S(\alpha + \beta) = S(\alpha) + S(\beta), \qquad S(c\alpha) = c \cdot S(\alpha)$$

(c any number in k).

Moreover

$$S(1) = n1.$$

The norm has the multiplicative property

$$\mathrm{Nm}(\alpha\beta) = \mathrm{Nm}\,\alpha \cdot \mathrm{Nm}\,\beta, \tag{1.7}$$

and for any number c in k

$$Nm(c) = c^n.$$

One is tempted to write the characteristic equation f(t) as Nm(t - α), and this is indeed justified if one replaces k and ϰ by the rings k[t], ϰ[t] of polynomials of t with coefficients in k and ϰ respectively. ϰ[t] is of degree n with respect to k[t], and the basis $\omega_1, \ldots, \omega_n$ of ϰ/k is also a basis for ϰ[t] relative to k[t]. Later on (§9) we shall study the adjunction of indeterminates more systematically. An equation like Nm(t - α) = f(t) stays true if one substitutes for t a number in k; substitution for t of an element in ϰ, however, is strictly forbidden!

One effective way to make use of ϰ being a division algebra (third level) is by the fact:

$$\alpha \neq 0 \text{ implies } Nm\,\alpha \neq 0.$$

Indeed, with the inverse α^{-1} of $\alpha \neq 0$, $\alpha\alpha^{-1} = 1$, one infers from (1.7):

$$Nm(\alpha) \cdot Nm(\alpha^{-1}) = 1.$$

The trace

$$S(\xi\eta)$$

is a symmetric bilinear form of the two variable elements ξ, η, which we call their scalar product. Expressing it in terms of a given basis ω_i,

$$\xi = \sum_i x_i\omega_i, \qquad \eta = \sum_k y_k\omega_k,$$

one gets

$$S(\xi\eta) = \Sigma s_{ik}x_i y_k \tag{1.8}$$

with the coefficients

$$s_{ik} = S(\omega_i\omega_k).$$

Our field is said to be <u>non-degenerate</u>, if this bilinear form is non-degenerate, i.e., if α = 0 is the only element for which the equation S(αη) = 0 holds identically in η.

This means that the discriminant of the basis ω_i, namely

$$D(\omega_1,\ldots,\omega_n) = \det(s_{ik})$$

is different from zero. Under the transformation (1.2) of the basis, the determinant of the invariant form (1.8) assumes the factor $|l_{ik}|^2$,

$$D(\omega_1^*,\ldots,\omega_n^*) = |l_{ik}|^2 \cdot D(\omega_1,\ldots,\omega_n).$$

In a non-degenerate field the discriminant of no basis vanishes while in a degenerate field the discriminant of every basis vanishes.

If $\alpha \neq 0$ one has

$$S(\alpha\xi) = n1 \quad \text{for} \quad \xi = \alpha^{-1}.$$

Hence our field $\varkappa$ is certainly non-degenerate unless $n1 = 0$, i.e., unless k is of a prime characteristic dividing the degree $[\varkappa:k]$.

2. Tower. Analysis of the Field Equation

"Finite field (over k)" is used as a shorthand term to describe fields of finite degree over k.

Theorem I 2, A. *If $\varkappa$ is a finite field (over k) of degree n and K a finite field over $\varkappa$ of degree r, then K is a finite field (over k) of degree $N = n \cdot r$:*

(2.1) $$[K:k] = [K:\varkappa] \cdot [\varkappa:k].$$

We speak of r as the relative degree of $K/\varkappa$ while n and N are the "absolute" degrees of $\varkappa$ and K respectively. k is considered as the ground level on which our structures rise, and "absolute" therefore means the same as "relative to k." At the present stage we apply the word "number" indiscriminately to the elements of k and of the fields over k. Using $\varkappa$ as ground field over which to erect a superstructure like K is an effective means of availing oneself of the commutative nature of the division algebra $\varkappa$.

The proof of our theorem is very simple indeed. Let Ω_s ($s = 1,\ldots, r$) be a relative basis of $K/\varkappa$ so that each number Ξ of K is uniquely expressible as

$$\Xi = \sum_{s=1}^{r} \xi_s\Omega_s \quad (\xi_s \text{ in } \varkappa),$$

and let $\omega_i (i = 1, \ldots, n)$ be a basis of $\varkappa$ in terms of which the coefficients ξ_s in their turn are expressible:

$$\xi_s = \sum_{i=1}^{n} x_{is}\omega_i.$$

One then realizes that the numbers

$$(2.2) \qquad \omega_i \Omega_s \ (i = 1, \ldots, n; \ s = 1, \ldots, r)$$

constitute an absolute basis for K and thus arrives at our fundamental law (2.1): Great degree = small degree x relative degree.

One can iterate the process and erect an h-story tower of fields

$$k_0 = k, k_1, k_2, \ldots, k_h$$

over k where each story k_j is a field of finite degree n_j over the next lower one $k_{j-1} (j = 1, \ldots, h)$. The absolute degree of $k_h = \varkappa$ will be the product $n_1 \ldots n_h$. The heights of the several stories in a building add up to the total height of the building; the degrees multiply rather than add. Another simile we sometimes use is that of a telescope. In particular we refer to a basis built up according to (2.2) as a telescopic basis.

We return to the two-story tower k, $\varkappa$, K. A number α of $\varkappa$ has its field equation $f_\varkappa(t)$ of degree n; but α is at the same time a number in K and as such has a field equation $f_K(t)$ in K. The coefficients of both polynomials $f_\varkappa(t)$ and $f_K(t)$ lie in the ground field k. I maintain

$$(2.3) \qquad f_K(t) = \{f_\varkappa(t)\}^r.$$

Included in this equation are the following relations for trace and norm:

$$(2.4) \qquad S_K(\alpha) = r \cdot S_\varkappa(\alpha), \qquad Nm_K(\alpha) = \{Nm_\varkappa(\alpha)\}^r.$$

Indeed, the matrix $A = \|a_{ik}\|$ associated with α in the regular representation in $\varkappa$ is determined by the equations (1.5) which imply

$$\alpha \cdot \omega_i \Omega_s = \sum_k a_{ki} \cdot \omega_k \Omega_s.$$

Hence in using the basis (2.2) for K one sees that the

representing matrix in K is

$$\begin{Vmatrix} A & 0 & \dots & 0 \\ 0 & A & \dots & 0 \\ \cdot & \cdot & \cdot & \cdot \\ 0 & 0 & \dots & A \end{Vmatrix} \qquad \text{(r rows)}$$

whence (2.3) and (2.4) follow immediately.

We make use of (2.3) for the analysis of the field equation, proving that the field equation is a power of the irreducible equation in k which is satisfied by α. We now again deal with a single field $\varkappa$ over k of which α is an element. Let m be the least exponent such that

$$1, \alpha, \alpha^2, \dots, \alpha^m$$

are linearly dependent in k,

$$\alpha^m + b_1\alpha^{m-1} + \dots + b_m = 0.$$

We have normalized as 1 the highest coefficient b_0 since b_0 is certainly $\neq 0$. The polynomial

$$g(t) = t^m + b_1t^{m-1} + \dots + b_m$$

is irreducible in k; for if it could be split, α would be a root of one of the factors which is of lower degree than g. Hence any polynomial q(t) in k is either divisible by g(t) or prime to g(t). In the latter case there exists another polynomial $q^*(t)$ in k such that

$$(2.5) \qquad q(t) \cdot q^*(t) \equiv 1 \pmod{g(t)}.$$

Consequently this case is impossible if $q(\alpha) = 0$; in other words, any polynomial q(t) which vanishes for $t = \alpha$ is divisible by g(t). The numbers in $\varkappa$ which are expressible by α in an integral rational manner, i.e., the numbers of the form $q(\alpha)$ where q(t) is any polynomial in k, not only form a ring, but even a field $k(\alpha)$. Indeed, if $\beta = q(\alpha) \neq 0$, then q(t) is prime to g(t) and (2.5) yields an inverse β^{-1} of β in the form $\beta^{-1} = q^*(\alpha)$. The field $k(\alpha)$ is of degree m and

$$1, \alpha, \alpha^2, \dots, \alpha^{m-1}$$

constitute a basis for it (natural basis). Any number in $k(\alpha)$ is uniquely expressible as a polynomial

$$c_0 + c_1\alpha + \dots + c_{m-1}\alpha^{m-1}$$

in α of formal degree $m - 1$ with coefficients c_i in k.

The field equation of α in $k(\alpha)$ is $g(t)$. What else could it be? It must be a polynomial divisible by $g(t)$ and of the same degree m. One can readily confirm this by direct calculation. In using the natural basis for $k(\alpha)$ one obtains from

$$\begin{aligned} &\alpha \cdot 1 = \alpha, \\ &\alpha \cdot \alpha = \alpha^2, \\ &\quad \cdots \\ &\alpha \cdot \alpha^{m-1} = -b_m - \dots - b_1\alpha^{m-1} \end{aligned}$$

as the sought-for polynomial

$$\begin{vmatrix} t, & 0, & 0, & \dots, & b_m \\ -1, & t, & 0, & \dots, & b_{m-1} \\ 0, & -1, & t, & \dots, & b_{m-2} \\ \cdot & \cdot & \cdot & \cdots & \cdot \\ 0, & 0, & 0, & \dots, & t + b_1 \end{vmatrix}$$

In replacing the first row by the linear combination of the first, second, ..., m^{th} rows with the coefficients $1, t, \dots, t^{m-1}$, one arrives at the result wanted.

$\varkappa$ must be a field over $k(\alpha)$ of a certain relative degree r, and by applying our former results to the tower $k \subset k(\alpha) \subset \varkappa$ we find $n = m \cdot r$ and for the field equation $f(t)$ of α in k:

$$f(t) = \{g(t)\}^r.$$

If and only if f itself is irreducible, $\varkappa$ coincides with the field $k(\alpha)$. α is then said to be a determining or a primitive number of $\varkappa$.

3. Simple Extension

In the classical theory of algebraic equations one starts with a given polynomial $f(x)$ in k and asks for a <u>root</u> θ of the equation $f(\theta) = 0$. One can then construct the field $\varkappa = k(\theta)$ over k of which θ is a determining number. If f is irreducible and of degree n, then the field $\varkappa$

is of degree n. In the theory of equations, transition from θ to any number

$$\gamma = c_0 + c_1\theta + \dots + c_{n-1}\theta^{n-1} \quad (c_i \text{ in } k)$$

of $\varkappa$ is called Tschirnhausen transformation. In general, namely except for singular sets of values of the coefficients c_i, γ will again be a determining number. In replacing the equation $f(x) = 0$ by the field $k(\theta)$, a step which initiated the modern approach to algebra, we focus our attention automatically on those features of an equation which are invariant under Tschirnhausen transformations.

The most important example of a ground field is the field of common rational numbers for which I use the freely invented symbol $\wp$. If we operate in the field Ω of all complex numbers, then the so-called fundamental theorem of algebra asserts that every equation $f(x)$ in $\wp$ has a root θ in Ω, and thus we can form the simple extension $\wp(\theta)$. This algebraic number field is cut out from the continuum Ω. The standpoint thus described is that of analysis; however, it is also adopted without discussion in Hilbert's classical Zahlbericht. Our modern algebraists have accustomed us to a more abstract viewpoint.

Indeed it is possible to create the field $k(\theta)$ out of the given field k by a purely algebraic construction without resorting to a given embedding field like Ω in which $f(x) = 0$ has a solution. The procedure is due to Kronecker, but was long before him applied by Cauchy to the case:

$$k = \text{field } \Lambda \text{ of all real numbers}, \quad f(x) = x^2 + 1,$$

with the purpose of founding the calculus of imaginary quantities on a sound basis. Again we assume k to be an arbitrary field, and $f(x)$ to be an irreducible polynomial in k of degree n. We consider the ring of all polynomials $q(\theta)$ in k of an indeterminate θ with the convention that two polynomials are identified if they are congruent mod $f(\theta)$. The use of the letter θ instead of x shall indicate this convention. Owing to the assumption of the irreducibility of $f(x)$, the elements thus introduced not only form a ring but a field $\varkappa = k(\theta)$ over k of which

$$1, \theta, \theta^2, \dots, \theta^{n-1}$$

is a natural basis. In order to verify this, one simply has to repeat the argument about $k(\alpha)$ as given in the previous section. The equation $f(\theta) = 0$ holds good in $\varkappa$. Since this process obviously is uniquely determined by the polynomial $f(x)$ we may speak of adjunction of "the" root θ of f (not of "one of the roots" of f).

The argument to which we have just referred shows even more, namely this: if K is a field over k in which $f(x) = 0$ has a solution $x = \Theta$ then the subfield $k(\Theta)$ contained in K is isomorphic to the abstractly constructed field $\varkappa$. It is indeed the irreducibility of f which insures both the field character and the uniqueness of $\varkappa$ in the sense of isomorphism. An isomorphism is a one-to-one mapping $\alpha \rightleftarrows \alpha'$ which preserves the fundamental algebraic operations:

$$(\alpha + \beta)' = \alpha' + \beta', \quad (\alpha\beta)' = \alpha' \cdot \beta'.$$

An isomorphic mapping upon one another of two fields $\varkappa$ and $\varkappa'$ over or relative to k is one which leaves every number of k unaltered. We shall deal with such isomorphisms (and automorphisms) only.

The Kronecker construction is so valuable because by yielding fields rather than algebras it automatically takes account of divisibility and commutativity. If one wishes to make full use of these properties of the field, it seems natural, therefore, to build the field up as a tower of consecutive simple extensions.

In the field Ω of all complex numbers, a polynomial not only has a root, but splits into linear factors. This feature also can be emulated by a suitable algebraic construction. Again we start with a polynomial $f(x)$ in k which, however, this time is not assumed to be irreducible. Rather, let $g(x)$ be an irreducible factor of $f(x)$. Kronecker's construction yields a finite field $k(\theta_1) = k_1$ in which $g(x)$ has a root θ_1 and hence splits off the factor $x - \theta_1$. A fortiori,

$$f(x) = (x - \theta_1) \cdot f_1(x),$$

where $f_1(x)$ is a polynomial in k_1. By operating with k_1 and $f_1(x)$ in the same manner as with k and $f(x)$ before we obtain a finite field $k_2 = k_1(\theta_2)$ over k_1 in which

$$f_1(x) = (x - \theta_2) \cdot f_2(x).$$

When we keep up the same procedure, our consecutive extensions finally result in a field U (= universe) of the desired nature, in which $f(x)$ completely splits into linear factors $x - \theta_i$. Because we shall employ this field U only as a general container which gives us enough freedom of motion, without the nature or structure of U ever entering into our constructions, we here abstain from discussing the uniqueness of U.

Let us return to the case of an irreducible $f(x)$. The Kronecker field $\varkappa = k(\theta)$ then has n isomorphic images within U:

$$\theta \to \theta_1, \ldots, \theta \to \theta_n. \tag{3.1}$$

In order to describe an isomorphic mapping of $k(\theta)$ it suffices to exhibit the image θ' of θ, since $q(\theta)$ will then be mapped upon $q(\theta')$ [$q(t)$ being any polynomial in k]. We call these n isomorphisms (3.1) the conjugations of $\varkappa$ and the n copies $k(\theta_i)$ ($i = 1, \ldots, n$) the conjugate fields of $\varkappa$. It is not an infrequent occurrence that some of these fields coincide. For instance, in the case of a quadratic equation $x^2 - a$ the two conjugate fields $k(\sqrt{a})$ and $k(-\sqrt{a})$ obviously coincide, although the isomorphism $\sqrt{a} \to -\sqrt{a}$ is not the identity. But it may even happen, that some of the isomorphisms (3.1) themselves are identical, namely if the corresponding roots θ_i are equal.

We encounter here the question whether an irreducible polynomial $f(x)$ may have multiple roots. Such a root will also be a root of the derivative $df/dx = \dot{f}(x)$. $\dot{f}(x)$ will thus be divisible by $f(x)$, which is impossible on account of the lower degree of $\dot{f}(x)$ unless $\dot{f}(x)$ vanishes identically. Write

$$f(x) = a_0 + a_1x + \ldots + a_l x^l + \ldots .$$

Then

$$\dot{f}(x) = 1a_1 + 2a_2x + \ldots + la_l x^{l-1} + \ldots .$$

Our condition amounts to

$$la_l = 0 \quad (l = 1, 2, \ldots). \tag{3.2}$$

If k is of characteristic 0, this implies $a_l = 0$. Therefore

$$f(x) = \text{const.} = a_0$$

and thus the existence of an irreducible $f(x)$ with multiple roots is precluded in this case. However, if k is of prime characteristic p one can infer from (3.2) the equation $a_l = 0$ only for exponents l which are not multiples of p. $f(x)$ will then be a polynomial of x^p, and the roots α of a polynomial of this sort $f(x) = g(x^p)$ are in fact all p-fold in a field of characteristic p. Namely $f(\alpha) = 0$ implies $g(\alpha^p) = 0$, hence $g(y)$ contains the factor $y - \alpha^p$, and $f(x) = g(x^p)$ the factor

$$x^p - \alpha^p = (x - \alpha)^p.$$

The irreducible polynomial $f(x)$ is called separable and so is the corresponding extension $k(\theta)$ if the derivative $\dot{f}(x)$ does not vanish identically. In the case of a separable extension we have exactly n distinct isomorphisms.

Let us study a separable extension $\varkappa = k(\theta)$ of degree n. Any number α of $\varkappa$ has its n conjugates in the container U. I maintain that the field equation $f^*(t)$ of α splits in U into the linear factors

$$f^*(t) = (t - \alpha_1) \dots (t - \alpha_n). \tag{3.3}$$

This shows that in particular

$$\begin{aligned} S(\alpha) &= \alpha_1 + \dots + \alpha_n, \\ \mathrm{Nm}\,\alpha &= \alpha_1 \dots \alpha_n: \end{aligned} \tag{3.4}$$

the trace is the sum, the norm the product of the conjugates.

For any number

$$\xi = x_0 + x_1\theta + \dots + x_{n-1}\theta^{n-1} \quad (x_i \text{ in } k) \tag{3.5}$$

in $\varkappa$ the multiplication $\eta = \alpha \cdot \xi$ is expressed by

$$y_i = \sum_k a_{ik} x_k \quad (i, k = 0, 1, \dots, n - 1). \tag{3.6}$$

Passing to the conjugates we obtain from (3.5):

$$\xi_i = x_0 + x_1\theta_i + \dots + x_{n-1}\theta_i^{n-1}.$$

We now regard the x_i as variables. These equations then constitute a linear transformation of the variables x_i into new variables ξ_i with coefficients in U. The transformation is non-singular, or the Vandermonde determinant

$$(3.7) \qquad |1, \theta, \ldots, \theta^{n-1}|$$

(consisting of the n rows which arise from the one written down in replacing θ by each of its conjugates $\theta_1, \ldots, \theta_n$) is $\neq 0$, since the θ_i are supposed to be distinct. In terms of the new variables ξ_i the substitution (3.6) reads

$$\eta_1 = \alpha_1\xi_1, \ldots, \eta_n = \alpha_n\xi_n,$$

i.e., the linear mapping A is expressed by the diagonal matrix of the α_i, and thus its characteristic polynomial

$$= (t - \alpha_1) \ldots (t - \alpha_n).$$

We conclude this section by developing a formula for the discriminant

$$(3.8) \qquad D = D(1, \theta, \ldots, \theta^{n-1})$$

of the natural basis of $\varkappa$. The determinant (3.7) equals the difference product

$$\prod_{i>k}(\theta_i - \theta_k).$$

If one multiplies the determinant by itself, interchanging in the first factor rows and columns, one obtains as its square the discriminant D, namely the determinant whose (i,k)-element

$$= \theta_1^i \cdot \theta_1^k + \ldots + \theta_n^i \cdot \theta_n^k = S(\theta^i \cdot \theta^k)$$

$$[i,k = 0, 1, \ldots, n - 1].$$

Therefore

$$(3.9) \qquad D = |1, \theta, \ldots, \theta^{n-1}|^2$$

or

$$D = (-1)^{\frac{1}{2}n(n-1)} \cdot \prod_{i \neq k}(\theta_i - \theta_k).$$

This formula proves that the separable extension $\varkappa = k(\theta)$ is non-degenerate. From

$$f(x) = (x - \theta_1) \ldots (x - \theta_n)$$

one derives

$$\dot{f}(\theta_1) = (\theta_1 - \theta_2) \ldots (\theta_1 - \theta_n);$$

hence after introducing the number

$$\dot{f}(\theta) = \delta$$

which we call the differential ("Differente," Hilbert) of θ:

$$\prod_{i \neq k} (\theta_i - \theta_k) = \delta_1 \ldots \delta_n = \mathrm{Nm}\ \delta.$$

Our final result is that but for the sign $(-1)^{\frac{1}{2}n(n-1)}$ <u>the discriminant D is the norm of the differential</u> δ.

4. Relative Trace, Norm and Discriminant

The facts (3.4) and their proof are capable of an important generalization. Let $\varkappa = k(\theta)$ again be a separable extension, and K a finite field over $\varkappa$ of degree r. Notations as in §2, with this special choice of a basis for $\varkappa$:

$$(4.1) \qquad \omega_1 = 1, \quad \omega_2 = \theta, \ldots, \omega_n = \theta^{n-1}.$$

We consider a number Γ in K, its relative trace and norm

$$S_{K/\varkappa}(\Gamma) = \sigma, \qquad \mathrm{Nm}_{K/\varkappa}(\Gamma) = \gamma.$$

They are numbers in $\varkappa$. We maintain that the absolute trace $S = S_K$ and norm $\mathrm{Nm}_K = \mathrm{Nm}$ are given by the formulas

$$(4.2) \qquad S(\Gamma) = \sigma_1 + \ldots + \sigma_n, \qquad \mathrm{Nm}\ \Gamma = \gamma_1 \ldots \gamma_n$$

in terms of the conjugates of σ and γ in the container U of $\varkappa$. (3.4) is the special case $K = \varkappa$.

Any number Ξ in K can be expressed in terms of a given relative basis Ω_s,

$$(4.3) \qquad \Xi = \Sigma \xi_s \Omega_s \quad (\xi_s \text{ in } \varkappa;\ s = 1, \ldots, r).$$

The multiplication by Γ, $H = \Gamma \cdot \Xi$, then appears as a linear substitution of the components ξ_s of Ξ:

$$(4.4) \qquad \eta_s = \sum_t \gamma_{st} \xi_t \quad (s, t = 1, \ldots, r).$$

σ and γ are by definition trace and norm of the matrix $\|\gamma_{st}\|$. By expressing ξ_s (and η_s) in terms of the natural basis (4.1) of $\varkappa$,

$$(4.5)\qquad \xi_s = x_{s1}\omega_1 + \dots + x_{sn}\omega_n,$$

the equation (4.4) change into

$$y_{si} = \sum_{t,k} c_{si,tk}\cdot x_{tk} \quad \begin{bmatrix} s,t = 1,\dots, r \\ i,k = 1,\dots, n \end{bmatrix}.$$

and the matrix C of the c represents Γ relative to k. But from (4.4) there follows for the conjugates, which we now indicate by a superscript:

$$(4.6)\qquad \eta_s^{(i)} = \sum_t \gamma_{st}^{(i)}\cdot \xi_t^{(i)},$$

and from (4.5):

$$(4.7)\qquad \xi_s^{(i)} = x_{s1}\omega_1^{(i)} + \dots + x_{sn}\omega_n^{(i)}.$$

Again we look upon (4.7) as a non-singular linear transformation of the variables

$$x_{s1},\dots, x_{sn} \quad \text{into} \quad \xi_s^{(1)},\dots, \xi_s^{(n)}.$$

With respect to this new basis, the matrix of the mapping C breaks up according to (4.6), namely into the conjugate matrices

$$\left\| \begin{matrix} \|\gamma_{st}^{(1)}\| & 0 & \dots & 0 \\ 0 & \|\gamma_{st}^{(2)}\| & \dots & 0 \\ \cdot & \cdot & \cdot & \cdot \\ 0 & 0 & \dots & \|\gamma_{st}^{(n)} \end{matrix} \right\|.$$

This proves our statements. Moreover it establishes the following connection between the field equations F(t) and $\varphi(t)$ of Γ in K/k and K/$\varkappa$ respectively:

$$(4.8)\qquad F(t) = \varphi_1(t)\ \dots\ \varphi_n(t).$$

$\varphi(t)$ is a polynomial in $\varkappa$, and $\varphi_i(t)$ are the conjugate polynomials in U.

By combining these formulas with the special ones into which they turn for $K = \varkappa$ we arrive at important results which are independent of the container U and of the idea of conjugation:

Theorem I 4, A. *Great trace = small trace of relative trace,*

Great norm = small norm of relative norm,

Absolute field equation $F(t) = Nm_{\varkappa}(\varphi(t))$.

If we apply the formula (4.2) to the product of two arbitrary numbers

$$\Xi = \sum_u \xi_u \Omega_u, \quad H = \sum_v \eta_v \Omega_v \quad (u, v = 1, \ldots, r)$$

in K we find for their scalar product

$$S(\Xi H) = \sum_{uv} \sigma_{uv}^{(1)} \xi_u^{(1)} \eta_v^{(1)} + \ldots + \sum_{uv} \sigma_{uv}^{(n)} \xi_u^{(n)} \xi_v^{(n)}$$

where

$$\sigma_{uv} = \text{relative trace } (\Omega_u \Omega_v).$$

Through (4.7) this bilinear form will change into

$$\Sigma s_{ui,vk} \cdot x_{ui} y_{vk} \quad \left(\begin{matrix} u, v = 1, \ldots, r \\ i, k = 1, \ldots, n \end{matrix}\right).$$

Its determinant is the "great discriminant" of the telescopic basis (2.2) for K. The determinant of the substitution (4.7) for a fixed index s equals (3.7); hence the determinant of the whole substitution for $s = 1, \ldots, r$ equals its rth power. Since the determinant of a quadratic form takes on the square of the transformation determinant as a factor, we find

$$D = |1, \theta, \ldots, \theta^{n-1}|^{2r} \cdot |\sigma_{uv}^{(1)}| \ldots |\sigma_{uv}^{(n)}|.$$

$|\sigma_{uv}|$ is the relative discriminant $D_{K/\varkappa}(\Omega) = \partial$. If we set

$$|1, \theta, \ldots, \theta^{n-1}|^2 = d$$

we have

$$(4.9) \qquad D = d^r \cdot \partial_1 \ldots \partial_n = d^r \cdot Nm\, \partial$$

where d depends only on $\varkappa$. Choosing in particular $K = \varkappa$, $\Omega_1 = 1$, we realize that the small discriminant

$$D(1, \theta, \ldots, \theta^{n-1}) \text{ equals } d$$

in confirmation of (3.9). Hence this law:

Theorem I 4, B. *Great discriminant = (r^{th} power of small discriminant) times (small norm of relative discriminant).*

We mention this

Corollary: If K is non-degenerate relative to the separable extension $\varkappa$, then K is absolute non-degenerate.

If δ, but for the sign, is the relative norm of a number Δ (differential) in K, as will be the case if K is a separable extension of $\varkappa$, then our law assumes the simpler form that the absolute differential of K equals $\delta \cdot \Delta$ (small differential x relative differential), in the sense that D is the great norm of $\delta \cdot \Delta$. Hence the rule: the discriminant behaves like the norm of a "differential" which obeys the same multiplicative law as the degree.

5. Removal of the Hypothesis of Separability

We shall prove in this section by an entirely different approach that equation (4.8) holds, even without assuming $\varkappa = k(\theta)$ to be a separable extension. Let $\varphi(t)$ be the field equation of Γ in $K/\varkappa$. We form

$$\varphi_1(t) \ldots \varphi_n(t) = F(t)$$

and show that F(t) has all the properties that characterize it as the field equation of Γ in K/k; namely

1) F lies in k and has the correct degree $N = nr$.

2) Γ satisfies the equation $F(\Gamma) = 0$.

3) F(t) is power of an irreducible polynomial P(t) in k.

(1) will now be established by means of the theorem on symmetric functions. In expressing the coefficients of $\varphi(t)$ in terms of the natural basis of $\varkappa$ one obtains a polynomial of two variables g(t;u) in k such that

$$\varphi(t) = g(t;\theta).$$

One then has

$$F(t) = g(t;\theta_1) \dots g(t;\theta_n).$$

If we first take $\theta_1, \dots, \theta_n$ as independent variables, the coefficients of $F(t)$ appear as symmetric functions of $\theta_1, \dots, \theta_n$. They are expressible in an integral rational manner by the elementary symmetric functions of the same arguments, and for these one finally substitutes the numerical values as furnished by the defining equation

$$f(t) = (t - \theta_1) \dots (t - \theta_n).$$

One operates here first with n indeterminates $\theta_1, \dots, \theta_n$ while the substitution takes place in the container U of $\varkappa$.

(2) is not quite so trivial as it looks at first, because we have no common field in which $\theta_1, \dots, \theta_n$ and Γ lie. Divide $F(t)$ by $\varphi(t)$ in $\varkappa$ and denote the remainder of degree $< r$ by $\rho(t)$,

$$F(t) = \varphi(t) \cdot \lambda(t) + \rho(t).$$

Since $F(t)$ is divisible by $\varphi_1(t)$ in U, the corresponding equation for the first conjugate shows that $\rho_1(t)$ is divisible by $\varphi_1(t)$ in U, which is impossible on account of the degree of ρ_1 unless $\rho_1(t) = 0$. Hence $\rho(t) = 0$, and $\varphi(\Gamma) = 0$ entails $F(\Gamma) = 0$.

(3) We know that $\varphi(t)$ is a power of an irreducible polynomial $\varphi^*(t)$ in $\varkappa$. It is therefore sufficient to prove that $F(t)$ is a power of an irreducible polynomial $P(t)$ in k under the assumption that $\varphi(t)$ is any irreducible polynomial in $\varkappa$. This we prove as follows.

Let $P(t)$ by any irreducible factor of $F(t)$ in k. I assert that $P(t)$ is divisible by $\varphi(t)$ in $\varkappa$. Otherwise, since $\varphi(t)$ is irreducible in $\varkappa$, one could find two polynomials $\pi(t)$, $\psi(t)$ in $\varkappa$ such that

$$P(t)\pi(t) + \varphi(t)\psi(t) = 1.$$

One multiplies the corresponding n conjugate equations in U, and observing that

$$\varphi_1(t) \dots \varphi_n(c) \text{ is divisible by } P(t)$$

one arrives at an equation

$$P(t) \cdot \Pi(t) = 1$$

where $\Pi(t)$ lies in U. Such an equation is impossible.

This shows that $F(t)$ cannot contain two different irreducible factors $P(t)$ in k since both would have the common factor $\varphi(t)$ in $\varkappa$.

Comparison of the second and the last coefficients in the resulting equation (4.8) again yields the laws (4.2), in particular (3.3), (3.4). The relations of Theorem I 4, A thus hold irrespective of whether θ is separable or not. By means of the rule for the trace, we also reëstablish the relation (4.9) for the discriminant, in particular (3.9). We thus arrive at the sharper results:

Theorem I 5, A. *A simple extension is non-degenerate if and only if it is separable. K is non-degenerate if and only if $\varkappa = k(\theta)$ and $K/\varkappa$ are non-degenerate.*

If we construct a finite field $\varkappa$ over k by a succession of simple extension,

(5.1) $$k = k_0 \subset k_1 \subset \ldots \subset k_h = \varkappa, \quad k_j = k_{j-1}(\theta_j)$$

we see that $\varkappa$ is non-degenerate if and only if each of the extensions is separable.

After we have freed the fundamental laws about trace, norm and discriminant from the assumption of separability, it is even possible to substitute for $\varkappa$ any finite field over k. Let $\varkappa$ be generated by a succession (5.1) of simple extensions. We show by induction with respect to $j = 0, \ldots, h$ that

(5.2) $$S_K = S_{k_j}(S_{K/k_j})$$

for any finite field K over k_j. By inductive hypothesis we have

(5.3) $$S_K = S_{k_{j-1}}(S_{K/k_{j-1}});$$

moreover, since k_j arises from k_{j-1} by simple extension, with k_{j-1} as ground field,

(5.4) $$S_{K/k_{j-1}} = S_{k_j/k_{j-1}}(S_{K/k_j}).$$

If we substitute this in (5.3) and consider that (5.3) includes

$$S_{k_j} = S_{k_{j-1}}(S_{k_j/k_{j-1}})$$

we get the relation (5.2). Similarly for the norm and the discriminant. In the case of the discriminant we find

$$(5.5) \qquad D = d^r \cdot Nm_{\varkappa}(\partial), \qquad \partial = D_{K/\varkappa}(\Omega_1, \ldots, \Omega_r)$$

where d depends on $\varkappa$ only; $\varkappa$ is referred to the telescopic basis ω_i corresponding to the construction (5.1) of $\varkappa$, and D is the discriminant of the basis

$$\omega_i \Omega_s \qquad (i = 1, \ldots, n;\ s = 1, \ldots, r)$$

in K. From the special basis ω_i we might pass to an arbitrary basis by a linear transformation $\| l_{ik} \|$. As D then takes on the factor

$$|l_{ik}|^{2r},$$

the relation (5.5) in the form in which it was stated stays good even for the arbitrary basis. Afterward we realize by the specialization $K = \varkappa$, $\Omega_1 = 1$, that d is the small discriminant

$$D_{\varkappa}(\omega_1, \ldots, \omega_n)$$

and (5.5) results in the former law, Theorem I 4, B.

In view of the universal validity of the rules concerning traces, norms, and discriminants to which we finally have won through, one would think it possible to derive them by direct calculation in a few lines. However, such attempts have failed, apparently for the reason that there is no simple way of taking commutativity into account. Therefore our devious method which covers this feature by building up $\varkappa$ through consecutive adjunctions.

6. The Galois Case

A simple extension $\varkappa = k(\theta)$ of degree n over k cannot have more than n isomorphic mappings into (i.e., upon subfields of) a given field K (over k). Indeed, if $f(x) = 0$ is the determining equation for θ, the image θ' of θ must be a root in K of the same equation, and the image of any number $\alpha = q(\theta)$ in $\varkappa$ is uniquely determined by θ': $\alpha' = q(\theta')$. But $f(x) = 0$ cannot have more than n distinct roots in K.

$\varkappa$ is called a Galois field if it allows the maximum number n of automorphisms. An automorphism s: $\alpha \to \alpha^s$ is an isomorphic mapping of $\varkappa$ upon itself. Our definition implies (1) the separability of $\varkappa$ and (2) that in $\varkappa$ itself f(x) splits into n linear factors. The n automorphisms s form a group $\mathfrak{g}$, the Galois group of $\varkappa$.

The main content of the Galois theory is the one-to-one correspondence between the subfields $\varkappa'$ of $\varkappa$ and the subgroups $\mathfrak{g}'$ of $\mathfrak{g}$. Here are the essential theorems on which it is based.

Theorem I 6, A. *A number α invariant under all automorphisms lies in the ground field.*

An older proof of our proposition uses the average

$$\frac{1}{n} \cdot \sum_s \alpha^s$$

and observes that according to the theorem on symmetric functions this is a number in k for every number α in $\varkappa$. If $\alpha = \alpha^s$ for all s, the average is α itself. The argument fails, however, if k is of prime characteristic dividing n. The following procedure is better.

Let $\varkappa$ be of relative degree r over the field $k(\alpha)$ of degree m: $n = r \cdot m$. All the s are automorphisms of $\varkappa$ over $k(\alpha)$ and hence their number n must be $\leqq r$. This precludes any other possibility but $m = 1$.

Let $\varkappa'$ be any subfield of $\varkappa$ and consider the automorphisms s such that $\alpha^s = \alpha$ for all numbers α in $\varkappa'$. They form a subgroup $\mathfrak{g}' = \mathfrak{g}(\varkappa')$ which we associate with the subfield $\varkappa'$ of degree m. I maintain that $\varkappa/\varkappa'$ is Galois and $\mathfrak{g}'$ its group of automorphisms. Indeed, each element s of $\mathfrak{g}'$ obviously is a relative automorphism of $\varkappa/\varkappa'$ and vice versa; the very concept of a relative automorphism is that of an automorphism of $\varkappa$, leaving the numbers of $\varkappa'$ unaltered. Let r be the relative degree of $\varkappa/\varkappa'$ and θ satisfy the (irreducible) equation

$$\varphi(x) = x^r + \alpha_1 x^{r-1} + \ldots + \alpha_r = 0 \tag{6.1}$$

in $\varkappa'$ (i.e., the coefficients α_j lie in $\varkappa'$). Any relative automorphism s must carry θ into a root of the same equation (6.1). Since $\varphi(x)$ is a divisor of $f(x)$ this equation has exactly r roots in $\varkappa$ which are among the n roots of $f(x)$. θ' being any of them, $\theta \to \theta'$ defines an automorphism of $\varkappa/\varkappa'$--in the same manner as described before relative to k, namely by letting $\psi(\theta')$ correspond to $\psi(\theta)$, if $\psi(x)$ is any polynomial in $\varkappa'$. Hence the order of $\mathfrak{g}'$ equals the relative degree r and is connected with the degree m of the corresponding subfield by the equation $n = r \cdot m$.

We now are able to generalize the previous theorem:

Theorem I 6, B. *Any number α invariant under the substitutions of the group $\mathfrak{g}' = \mathfrak{g}(\varkappa')$ lies in $\varkappa'$.*

Indeed, form the field $\varkappa'(\alpha)$. Its associated group $\mathfrak{g}'$ is the same (it has not decreased); consequently its degree is the same as that of $\varkappa'$. Thus the relative degree $[\varkappa'(\alpha) : \varkappa']$ must be 1, or α lies in $\varkappa'$.

With any given subgroup $\mathfrak{g}'$ of $\mathfrak{g}$ we can associate the field $\varkappa' = k(\mathfrak{g}')$ of those numbers α of $\varkappa$ which satisfy the relation $\alpha^s = \alpha$ for all elements s of $\mathfrak{g}'$. What we have proved so far is the statement

$$k(\mathfrak{g}(\varkappa')) = \varkappa'.$$

It constitutes the first part of Galois' theory. In the second part we start with a given subgroup $\mathfrak{g}'$ of $\mathfrak{g}$ and prove

$$\mathfrak{g}(k(\mathfrak{g}')) = \mathfrak{g}'.$$

In doing so we must in some way make use of the fact that $\mathfrak{g}'$ is a subgroup rather than an arbitrary subset of $\mathfrak{g}$. We form the polynomial

$$(6.2) \quad \prod_{s \text{ in } \mathfrak{g}'} (x - \theta^s) = \varphi(x) = x^r + \alpha_1 x^{r-1} + \ldots + \alpha_r.$$

This polynomial and its coefficients are invariant under the substitutions s of $\mathfrak{g}'$,

$$\varphi^s(x) = \varphi(x) \qquad (x \text{ in } \mathfrak{g}').$$

Indeed,

$$\varphi^s(x) = \prod_{s' \text{ in } \mathfrak{g}'} (x - \theta^{s's}),$$

and s's, like s' itself, runs over all elements of $\mathfrak{g}'$. [$\alpha^{s's}$ stands for $(\alpha^{s'})^s$.] Consequently $\varphi(x)$ lies in $\kappa' = k(\mathfrak{g}')$. Vice versa, a substitution s leaving $\alpha_1, \ldots, \alpha_r$ unchanged, must carry θ into one of the roots of the equation $\varphi(x) = 0$, (6.2), and therefore coincide with one of the elements s of $\mathfrak{g}'$. This proves our theorem, yielding at the same time the further result that κ' arises from k by the simultaneous adjunction of $\alpha_1, \ldots, \alpha_r$:

$$\kappa' = k(\alpha_1, \ldots, \alpha_r).$$

The n automorphisms s of κ define m conjugations, i.e., isomorphisms of κ' upon certain conjugate subfields in κ. This is so because the substitutions s of the same coset modulo $\mathfrak{g}'$,

$$s's \qquad (s' \text{ ranging over } \mathfrak{g}')$$

yield the same conjugation of κ'.

The study of the Galois group $\mathfrak{g}$ of κ results in a complete survey of all subfields of κ: Each subgroup $\mathfrak{g}'$ of $\mathfrak{g}$ defines such a field consisting of the numbers invariant under the substitutions of $\mathfrak{g}'$; and this relationship is one-to-one. The larger the group, the smaller the corresponding field. We see in particular that κ has but a finite number of subfields.

The above contains the most important and pleasant part of the Galois theory. Its application in this form is limited by the fact that the subfields $k(\alpha_1, \ldots, \alpha_r)$ constructed by means of the subgroups, as far as we can tell now, are of a more general nature than the original field κ, inasmuch as they appear not to spring each from a single determining number. The next section will be devoted to a discussion of this problem.

7. Consecutive Extensions Replaced by a Single One

For s any substitution of the Galois group, $\varphi^s(x)$ has the root θ^s. Since $\varphi^{s's}(x) = \varphi^s(x)$ if s' is in $\mathfrak{g}'$, $\varphi^s(x)$ has also the root $\theta^{s's}$, and therefore

$$\varphi^s(x) = \prod_{s' \text{ in } \mathfrak{g}'} (x - \theta^{s's}).$$

The m polynomials thus obtained,

$$\varphi_1(x), \ldots, \varphi_m(x),$$

which correspond to the m cosets $\mathfrak{g}/\mathfrak{g}'$ are different because their roots differ. If k is not strictly finite, i.e., if it contains infinitely many numbers, then we can choose a number c in k such that for x = c the values of $\varphi_2(x), \ldots, \varphi_m(x)$ are different from the value of $\varphi_1(x)$; one simply chooses c so as not to annul the polynomial

$$(\varphi_2(x) - \varphi_1(x)) \ldots (\varphi_m(x) - \varphi_1(x))$$

which has at most $(m-1)(r-1)$ roots in k. We set $\varphi(c) = \eta$. The only s satisfying the condition $\eta^s = \eta$ are those in $\mathfrak{g}'$, and therefore

$$\varkappa' = k(\eta).$$

Hence we have found a determining number η for the arbitrary subfield $\varkappa'$.

The case of a strictly finite k will be treated separately in the next section.

We can now state that $\varkappa'$ is itself a Galois field if and only if $\mathfrak{g}'$ is an invariant subgroup of $\mathfrak{g}$, the factor $\mathfrak{g}/\mathfrak{g}'$ being the Galois group of $\varkappa'$.

We turn next to a general study of consecutive extensions of a field k which is not strictly finite.

$$\varkappa = k(\theta), \qquad K = \varkappa(\Theta).$$

We suppose the first extension of degree n to be separable. Let f(x), g(x) be the field equations in k for θ and Θ respectively;

$$f(x) = (x - \theta) \cdot f^*(x)$$

We construct a finite field U over K wherein $f^*(x)$ splits into linear factors. We then have in U:

$$f(x) = (x - \theta_1)(x - \theta_2) \ldots (x - \theta_n) \qquad (\theta_1 = \theta).$$

g(x) will have a certain number of distinct roots $\Theta_1 = \Theta, \ldots, \Theta_M$ in U. We maintain that $K = k(H)$ where H is the linear combination

$$H = \Theta + c\theta$$

with a suitable coefficient $c \neq 0$ in k. Indeed

(7.1) $$g(H - c\theta) = 0, \qquad f(\theta) = 0.$$

We look upon the first equation as an algebraic equation in k(H) for θ. The two equations have a common solution $\theta = \theta_1$. Can they have more than one, i.e., is it possible that one of the other roots $\theta_2, \ldots, \theta_n$ of f also satisfies the first equation? Only if for a certain pair (i,l)

$$H - c\theta_i = \Theta_l \qquad (i = 2, \ldots, n;\ l = 1, \ldots, M)$$

or

$$c(\theta_1 - \theta_i) = \Theta_l - \Theta_1.$$

What we shall do then is to choose c as a number in k, different from 0 and the (n - 1)(M - 1) quotients

$$\frac{\Theta_l - \Theta_1}{\theta_1 - \theta_i} \quad \left(\begin{matrix} i = 2, \ldots, n \\ l = 2, \ldots, M \end{matrix}\right).$$

Then (7.1) have but the one root $\theta_1 = \theta$ in common which may be isolated by applying the Euclidean algorithm of greatest common divisor to the polynomials

$$g(H - cx) \quad \text{and} \quad f(x).$$

Consequently θ and $\Theta = H - c\theta$ lie in k(H), or K = k(H). The two consecutive extensions θ, Θ may be replaced by the simple adjunction of H to k.

The case of a strictly finite ground field will be settled in the next section.

In particular, we may now be sure that every non-degenerate field has a determining number.

Once more we return to the Galois theory in order to round out the preceding results, taking the viewpoint from which Galois himself started. Let f(x) be a polynomial of degree n in k which is prime to its derivative. In §3 we have constructed a "universe" U in which f(x) splits into n linear distinct factors,

$$f(x) = (x - \theta_1)(x - \theta_2) \ldots (x - \theta_n).$$

We maintain that this construction yields a Galois field U.

Proof: Since U arises by a chain of separable adjunctions, U itself possesses a determining number Θ,

$$U = k(\Theta).$$

Let the consecutive adjunctions in the chain

$$k = k_0 \subset k_1 \subset \ldots \subset k_h = U$$

be of degrees $n_1, \ldots, n_h$,

$$n_i = [k_i : k_{i-1}].$$

The degree N of U equals $n_1 \ldots n_h$. The field k_1 arises from k_0 by adjoining the root θ_1 of an irreducible factor $f_1(x)$ of $f(x)$ in k_0. In U we shall have

$$f_1(x) = (x - \theta)(x - \theta')\ldots$$

where $\theta = \theta_1$, and $\theta', \ldots$ are some of the roots $\theta_2, \ldots, \theta_n$. By

$$\theta_1 \to \theta, \qquad \theta_1 \to \theta', \ldots$$

one defines n_1 conjugations of k_1 within U. By induction with respect to i we proceed to ascertain $n_1 \ldots n_i$ conjugations of k_i within U. Let $\varphi(x) = 0$ of degree $n_i = r$ be the determining equation of k_i/k_{i-1},

$$\varphi(\theta_i) = 0, \qquad k_i = k_{i-1}(\theta_i),$$

and $\alpha \to \alpha'$ be any of the $n_1 \ldots n_{i-1} = \bar{n}$ conjugations of k_{i-1} within U. We are able to extend this conjugation in r different ways into k_i, so that the $\bar{n}$ conjugations of k_{i-1} lead to $\bar{n} \cdot r = n_1 \ldots n_i$ conjugations of k_i. In fact $\varphi(x)$ is a divisor of the polynomial $f(x)$, and so is its conjugate $\varphi'(x)$. Hence $\varphi'(x) = 0$ has r roots $\theta_1', \ldots, \theta_r'$ in U which are among the n numbers $\theta_1, \ldots, \theta_n$. Any conjugation of k_i which coincides with the given conjugation $\alpha \to \alpha'$ in k_{i-1} must send θ_i into one of the r numbers $\theta_1', \ldots, \theta_r'$. Vice versa, the given conjugation of k_{i-1} together with

$$\theta_i \to \theta_1' \text{ or } \theta_i \to \theta_2' \ldots \text{ or } \theta_i \to \theta_r'$$

determines a conjugation of k_i. Indeed, any number in k_i equals $\psi(\theta_i)$ where ψ denotes a polynomial in k_{i-1}. In sticking to the first choice $\theta_i \to \theta_1'$ the conjugation of k_i sought-for is set up by the rule

$$\psi(\theta_i) \to \psi'(\theta_1')$$

which is unambiguous because

$$\psi(\theta_i) = \psi_*(\theta_i) \quad \text{or} \quad \psi(x) \equiv \psi_*(x) \pmod{\varphi(x)}$$

($\psi(x)$ and $\psi_*(x)$ polynomials in k_{i-1}) implies

$$\psi'(x) \equiv \psi'_*(x) \pmod{\varphi'(x)}.$$

This inductive construction finally yields N conjugations of U within U, i.e., N automorphisms of U. Each of them will be described by a permutation of the n roots $\theta_1, \ldots, \theta_n$, so that the Galois group of U appears as a group of permutations in n figures, in agreement with Galois' original conception.

The same construction of extending an isomorphism shows that any field in which $f(x)$ decomposes into linear factors contains a subfield isomorphic with the above Galois field $U = k(\theta_1, \ldots, \theta_n)$. Hence the structure of the (minimum) embedding Galois field is uniquely determined.

What we have proved may also be stated thus: Given several non-degenerate fields $k_1, \ldots, k_m$, one can find a Galois field K which contains isomorphic images of each of the given fields $k_1, \ldots, k_m$.

8. Strictly Finite Fields

The situation is much simpler if k is strictly finite. (The strictly finite fields were first investigated by Galois, and therefore they are also often called by his name; our term "strictly finite" will prevent possible confusions.)

Theorem I 8, A. *Any field $\varkappa$ of finite degree over a strictly finite field k is a Galois field whose Galois group is cyclic.*

A strictly finite field k is necessarily of prime characteristic p and of finite degree f over the absolute ground field $\mathfrak{g}_p$ of characteristic p which consists of the ordinary integers mod p. If $\omega_1, \ldots, \omega_f$ is a basis of $k/\mathfrak{g}_p$ the expression

$$a_1\omega_1 + \ldots + a_f\omega_f \quad (a_i \text{ in } \mathfrak{g}_p)$$

for the generic element of k shows that k contains $P = p^f$ elements. Hence the number of elements in a strictly finite field always is a prime power.

The elements $\neq 0$ form a multiplicative group k^* of degree $P - 1$. Hence every such element α satisfies the equation

$$\alpha^{P-1} = 1,$$

and every element of k whatsoever (0 not excluded) satisfies the equation

$$\alpha^P = \alpha.$$

This could also be proved in a manner similar to the common proof of Fermat's theorem $a^p = a$ in $\mathfrak{g}_p$; and the familiar methods for constructing a primitive residue mod p likewise carry over to the present case. We shall thus find an element ρ of k^* such that every element α of k^* is a power of ρ. We repeat the argument.

In the multiplicative group k^* let e be the order of α. The order e will be a divisor of the degree $P - 1$ of the group. The equation $\xi^e = 1$ will have the e distinct solutions

$$1, \alpha, \ldots, \alpha^{e-1}$$

and no others, because the number of distinct roots cannot exceed e. Such a power α^i is of the exact order e, if and only if i is prime to e. Hence the number $\psi(e)$ of solutions of the equation $\xi^e = 1$ is either zero or the Euler function $\varphi(e)$. The fact that every element of our group has a definite order is expressed by the equation

$$\sum_{e|P-1} \psi(e) = P - 1.$$

Comparing this with Euler's equation

$$\sum_{e|P-1} \varphi(e) = P - 1$$

one realizes that in

$$\psi(e) = 0 \quad \text{or} \quad \varphi(e)$$

the second alternative always prevails. Consequently there exist $\varphi(P - 1)$ elements of order $P - 1$. Any one of these "primitive roots" ρ reveals the group under consideration to be cyclic.

Let now $\varkappa$ be a field of finite degree n over k. It will consist of P^n elements. A "primitive root" of $\varkappa$ is a determining number of $\varkappa$ over k (even over $\mathfrak{g}_p$). The equations

$$(\alpha + \beta)^P = \alpha^P + \beta^P, \qquad (\alpha\beta)^P = \alpha^P \cdot \beta^P$$

together with

$$a^P = a \qquad \text{(for a in k)}$$

show $\alpha \to \alpha^P$ to be an automorphism of $\varkappa/k$. Thus we obtain the n distinct automorphisms of $\varkappa$ relative to k:

$$\rho \to \rho, \quad \rho \to \rho^P, \quad \rho \to \rho^{P^2}, \ldots, \rho \to \rho^{P^{n-1}}.$$

Hence $\varkappa/k$ is a Galois field whose Galois group is cyclic with the generating element $\rho \to \rho^P$.

This analysis of strictly finite fields will prove important for the arithmetical theory of residues in arbitary number fields.

9. Adjunction of Indeterminates

We close with a systematic study of a very elementary subject of which casual application has been made before, namely the adjunction of indeterminates.

The polynomials $\varphi(x,y,\ldots)$ of a given number of indeterminates x,y,... with coefficients in a given field k form a ring k x,y,... without null divisors. If $\omega_1,\ldots,\omega_n$ is a basis of $\varkappa$ over k, then $\varkappa[x,y,\ldots]$ evidently has the same basis over $k[x,y,\ldots]$ and is therefore of the same degree n.

The field equation of the polynomial $\varphi(xy..)$ in $\varkappa[xy..]$ over $k[xy..]$ will be

$$Nm\left\{t - \varphi(xy..)\right\} = t^n - f_1t^{n-1} + \ldots \pm f_n$$

where the coefficients f_i lie in $k[xy..]$. By expressing the fact that this polynomial of t vanishes after the substitution $t = \varphi(xy..)$ we get the equation

$$f_n = Nm\ \varphi(xy..)$$

$$= \varphi(xy..) \cdot \left\{f_{n-1} - f_{n-2}t + \ldots \pm t^{n-1}\right\}_{t\,=\,\varphi(xy..)},$$

proving the important principle that the norm of $\varphi(xy..)$ contains $\varphi(xy..)$ itself as a factor. We often write

$$\mathrm{Nm}\ \varphi(xy..) = \varphi(xy..) \cdot T_\varphi(xy..)$$

and call φ and T_φ head and tail of the norm.

It is important to observe that

$$\varphi \neq 0 \quad \text{implies} \quad \mathrm{Nm}\ \varphi \neq 0.$$

The proof for polynomials essentially differs from that for numbers. Let us arrange φ according to powers of x,

$$\varphi(xy..) = \varphi_0 x^g + \varphi_1 x^{g-1} + \ldots + \varphi_g$$

where $\varphi_0, \ldots, \varphi_g$ depend on $y, \ldots$ only and $\varphi_0 \neq 0$. Denoting the numbers of indeterminates $x, y, \ldots$ by ν, we assume our statement which holds good for $\nu = 0$ to be true for $\nu - 1$ arguments. The very definition of the norm shows that

$$\mathrm{Nm}\ \varphi(x,y,..) = \mathrm{Nm}\ \varphi_0 \cdot x^{gn} + \ldots,$$

and hence we have the implications

$$\varphi_0 \neq 0 \longrightarrow \mathrm{Nm}\ \varphi_0 \neq 0 \longrightarrow \mathrm{Nm}\ \varphi \neq 0.$$

From the ring $\varkappa[xy..]$ we pass to the field $\varkappa(xy..)$ of rational functions

$$\alpha = \frac{\varphi(xy..)}{\psi(xy..)}$$

in $\varkappa$. Numerator and denominator are polynomials in $\varkappa$, and the latter, ψ, $\neq 0$.

$$\mathrm{Nm}\ \psi = \psi \cdot T_\psi = \psi \cdot \tau$$

is an element $d(xy..) \neq 0$ of $k(xy..)$, and if we write

$$\alpha = \frac{\varphi\tau}{d}, \qquad \varphi\tau = a_1\omega_1 + \ldots + a_n\omega_n,$$

by representing the polynomial $\varphi\tau$ in $\varkappa$ in terms of the basis $\omega_1, \ldots, \omega_n$ of $\varkappa/k$, we see that $\omega_1, \ldots, \omega_n$ is also a basis and n the degree of

$$\varkappa(xy..)/k(xy..).$$

The field equation of α equals

$$Nm(t - \alpha) = \frac{Nm(t\psi - \varphi)}{Nm\ \psi}.$$

It is a power of the irreducible polynomial $f(t)$ in $k(xy..)$ of which α is a root.

Chapter II

THEORY OF DIVISIBILITY (KRONECKER, DEDEKIND)

1. Integers

The integers in the field $\mathfrak{r}$ of rational numbers form a ring. We face the problem of selecting from the numbers of a given field k the integers in such a way that they form a ring [k] containing the unit 1. Moreover we require that k be the quotient field of [k], i.e., that any number in k may be written as a fraction a/b of integers a and b with b $\neq$ 0. Let us suppose this has in some way been accomplished so that we know what an integer in k is. We pass to a finite field $\varkappa$ over k and wish to extend the definition of integers to $\varkappa$ in such a way that the integers of k stay integral in $\varkappa$ and the above requirements are also satisfied in $\varkappa$. For this "carry over"-problem there exists a universal solution.

<u>Definition</u>: *The number α in $\varkappa$ is said to be integral (in $\varkappa$) if satisfying an equation*

$$g(\alpha) = \alpha^h + a_1\alpha^{h-1} + \ldots + a_h = 0$$

with integral coefficients $a_1, \ldots, a_h$ in k.

Clearly the integers in k are integers in this new sense. Take a second integer β with the equation

$$\bar{g}(\beta) = \beta^l + b_1\beta^{l-1} + \ldots + b_l = 0 \quad (b_i \text{ integers in k}).$$

We want to prove that $\alpha \pm \beta$ and $\alpha\beta$ are also integers. For this purpose we arrange in a single file the hl numbers

$$\omega_j = \alpha^i\beta^m \quad (i = 0, \ldots, h - 1; m = 0, \ldots, l - 1)$$

and use them, as we have before used the basis, for the construction of an algebraic equation for $\gamma = \alpha + \beta$. We have

$$\gamma\omega_j = \alpha^{i+1}\beta^m + \alpha^i\beta^{m+1}.$$

Both terms at the right side are ω_j's themselves, provided $i < h - 1$ and $m < l - 1$. However, if $i = h - 1$ the first term

$$\alpha^h\beta^m \quad \text{equals} \quad -a_1\alpha^{h-1}\beta^m - \dots -a_h\beta^m.$$

Similarly for the second term if $m = l - 1$. In any case $\gamma\omega_j$ is a linear combination

$$\gamma\omega_j = \sum_{j'} c_{j'j}\omega_{j'}$$

with integers $c_{j'j}$ in k. This yields an equation of degree hl for γ,

$$|\gamma\delta_{j'j} - c_{j'j}| = 0$$

whose coefficients arise from the $c_{j'j}$ by multiplication, addition and subtraction, and are therefore integers in k. The same method applies to $\alpha - \beta$ and $\alpha\beta$, and more generally to any integral polynomial $f(\alpha,\beta)$ in k, i.e., any polynomial whose coefficients are integers in k. One has merely to reduce the polynomials $f(\alpha,\beta)\omega_j$ of α and β modulis $g(\alpha)$ and $\overline{g}(\beta)$.

Any number α in $\varkappa$ may be written as a fraction of integers, even as a fraction whose denominator lies in k. Indeed, write the coefficients of the field equation of α as fraction in k with the denominators $d_1, \dots, d_n$. Putting $d = d_1 \dots d_n$ one gets

$$\alpha^n + \frac{a_1}{d}\alpha^{n-1} + \dots + \frac{a_n}{d} = 0 \quad (a_i \text{ integers}).$$

This shows that $d\alpha$ is integral in $\varkappa$:

$$(d\alpha)^n + a_1(d\alpha)^{n-1} + da_2(d\alpha)^{n-2} + \dots + d^{n-1}a_n = 0.$$

The following statement is important:

Theorem II 1, A. *If the number Γ in $\varkappa$ satisfies an equation*

$$\Gamma^r + \alpha_1\Gamma^{r-1} + \dots + \alpha_r = 0$$

whose coefficients α_i are integers in $\varkappa$, then Γ is an integer.

We prefer to prove it in the following form. Let there be given some integers $\alpha, \beta, ..$ in $\varkappa$ and integral polynomials

$$f_1(x,y,..), ..., f_r(x,y,..)$$

in k. A number Γ in $\varkappa$ satisfying the equation

$$\Gamma^r + f_1(\alpha,\beta,..)\Gamma^{r-1} + ... + f_r(\alpha,\beta,..) = 0$$

is necessarily integral. This form is preferable if one wishes not only to establish the existence of an equation for Γ with integral coefficients in k, but also to construct this equation in the most convenient way, in particular avoiding an undesirably high degree. We give the proof for two arguments α, β. With the same pseudo-basis ω_j as before we obtain

$$f_1(\alpha,\beta)\omega_j = \sum_{j'} c^{(1)}_{j'j}\,\omega_{j'}.$$

Hence

$$\sum_{j'}\left\{\Gamma^r\delta_{j'j} + \Gamma^{r-1}c^{(1)}_{j'j} + ... + c^{(r)}_{j'j}\right\}\omega_{j'} = 0,$$

and consequently the determinant

$$\left|\Gamma^r\delta_{j'j} + \Gamma^{r-1}c^{(1)}_{j'j} + ... + c^{(r)}_{j'j}\right|$$

vanishes. This is an equation for Γ of degree $r \cdot hl$ with integral coefficients from k.

It has been said that an integer in k is an integer in $\varkappa$; it has not been said and without further restrictions would not be true, that a number in k which is an integer in $\varkappa$ is also an integer in k.

2. Our Disbelief in Ideals

Let there be given a ring [k] of integers. When we ask with respect to a given integer $\delta \neq 0$ what numbers α are divisible by δ (for what α the quotient α/δ is integral), δ serves as divisor. Two different integers may give rise to the same divisor, inasmuch as any number divisible by the first integer is also divisible by the second one, and vice versa. On the other hand there exist classical examples, for instance the quadratic field $\mathfrak{R}(\sqrt{-5})$, which show that the law of unique decomposition of a divisor into prime divisors may not hold without some suitable extension of the notion of a divisor.

Whatever a divisor $\mathfrak{a}$ may be, it certainly will be characterized by the set of integers divisible by $\mathfrak{a}$; we shall require that the divisibility of α by $\mathfrak{a}$ implies the same for $\lambda\alpha$ whatever the integer λ, and divisibility of α and β implies divisibility of $\alpha \pm \beta$. In other words, the numbers divisible by $\mathfrak{a}$ form an <u>ideal</u>; in fact, an ideal is a subset of a given ring [k] of numbers having these two properties:

$\alpha \pm \beta$ belong to the ideal if α and β do;

$\lambda\alpha$ belongs to the ideal if λ is any number of the ring and α lies in the ideal.

(The set consisting of the one number 0 shall not be counted as an ideal.) The notion was first introduced by Dedekind for the arithmetics of algebraic number fields. Dedekind substituted for a divisor $\mathfrak{a}$ the ideal of the integers divisible by $\mathfrak{a}$, e.g., in the ring [$\mathfrak{g}$] of ordinary integers the divisor 3 by the ideal of numbers

$$\ldots, -9, -6, -3, 0, 3, 6, 9, \ldots$$

Instead of saying that α is divisible by the divisor $\mathfrak{a}$, he says that α is an element of the set or ideal $\mathfrak{a}$. I prefer to stick to the more suggestive divisor terminology.

An integer α serving as divisor is called the <u>principal</u> divisor (α); the principal ideal (α) arises by multiplying α with all integers λ. More generally,

$$(\alpha_1, \ldots, \alpha_r)$$

denotes the ideal $\mathfrak{a}$ consisting of all numbers of the form

$$\lambda_1\alpha_1 + \ldots + \lambda_r\alpha_r$$

(λ_i arbitrary integers), and $\alpha_1, \ldots, \alpha_r$ then form an (ideal) basis of $\mathfrak{a}$.

A priori there seems no reason for rejecting any ideal as representing a divisor, in the sense that the ideal consists of all numbers divisible by the divisor. Once this standpoint giving the notion of divisor its widest possible sense is adopted, product and greatest common divisor acquire an unambiguous meaning. Indeed, amongst all divisors $\mathfrak{c}$ which go into two given divisors $\mathfrak{a}$ and $\mathfrak{b}$, there is a greatest one $(\mathfrak{a}, \mathfrak{b})$ which is divisible by each

$\mathfrak{c}$; the corresponding ideal, the smallest one comprising all numbers of the two given ideals $\mathfrak{a}$ and $\mathfrak{b}$, consists of all numbers of the form

$$\alpha + \beta \quad (\alpha \text{ in } \mathfrak{a},\ \beta \text{ in } \mathfrak{b}).$$

The product $\mathfrak{a}\mathfrak{b}$ is the least ideal containing all products

$$\alpha\beta \quad (\alpha \text{ in } \mathfrak{a},\ \beta \text{ in } \mathfrak{b})$$

and therefore consists of all finite sums of such products. However, it should be emphasized that these definitions are legitimate only with respect to the totality of all ideals; their rights become challengeable as soon as the notion of divisor is narrowed down to a certain subclass of ideals.

When the ring [k] whose elements we called integers, is a field, the only ideal in [k] is [k] itself. For if $\alpha \neq 0$ lies in the given ideal $\mathfrak{a}$, so does any number of the form $\lambda\alpha$ and hence every number β in [k]; $\lambda = \beta/\alpha$. Every ideal $\mathfrak{a}$ is principal in the ring $[\mathfrak{g}]$ of common integers. For let a be the least positive element of $\mathfrak{a}$; then $\mathfrak{a} = (a)$. The same is true in the ring k[x] of all polynomials of a single variable x with coefficients taken from a given field k. Here a is to be taken as a non-vanishing element of the given ideal of lowest degree. The proposition breaks down in the ring $k[x_1, \ldots, x_m]$ of **k**-polynomials $f[x_1, \ldots, x_m]$ of several variables $(m > 1)$.

An equation $f(x_1 \ldots x_m) = 0$ represents an $(m - 1)$-dimensional algebraic surface in the m-dimensional space with the coördinates x_i, while a set of simultaneous equations

$$f_1(x_1 \ldots x_m) = 0, \ldots, f_r(x_1 \ldots x_m) = 0$$

defines an algebraic manifold M of lower dimensionality. On M there vanishes every polynomial of the form

$$L_1 f_1 + \ldots + L_r f_r$$

(L_i arbitrary polynomials) or of the ideal $(f_1, \ldots, f_r)$. Hence an algebraic manifold is defined by an ideal in the ring $k[x_1 \ldots x_m]$. Incidentally, by a famous theorem due to Hilbert, every ideal in $k[x_1 \ldots x_m]$ has a finite ideal basis. A point lies on the manifold if all the polynomials of the ideal vanish at the point. But the algebraic geometer distinguishes between the surface $f = 0$ and $f^2 = 0$;

i.e., what characterizes the manifold for him is not the set of points lying on M, but the defining ideal.

The law of unique decomposition holds good for the polynomials in m(>1) variables. Hence from the standpoint of a theory of divisibility there is no reason here for introducing "ideal factors" or divisors besides the elements of the ring themselves. On the contrary the law is irretrievably destroyed by passing from polynomials to polynomial ideals. Therefore when one widens the realm of elements to that of ideals in a given ring, one sometimes gains and sometimes loses. One gets the impression that, generally speaking, the truth lies halfway: if the domain of integers in many cases is too narrow, the domain of ideals is in most cases too wide. We admit that polynomial ideals are a worthy subject of study--not, however, as a tool for the arithmetic of polynomials, but for their own sake, because algebraic manifolds of lower dimension deserve no less attention than algebraic surfaces.

Our aim here is to secure the law of unique decomposition. With this sole purpose in mind we must reject Dedekind's notion of ideal as a universal solution. Rather, an axiomatic approach is urged upon us: after setting down in the next section our axioms including the law of unique decomposition, we shall endeavor to show that once these axioms are granted in the ground field k, one can extend the basic concepts of integers and divisors to any finite field over $\varkappa$ without invalidating the axioms. This is accomplished by following Kronecker's idea of adjoining indeterminates rather than by Dedekind's procedure.

3. The Axioms

We operate in a field k. Some of its numbers are distinguished as integers. The axioms are concerned with them in their relation to another class of objects, called divisors; the basic relation is divisibility of (an integer) a by (a divisor) $\mathfrak{m}$, in symbols a : $\mathfrak{m}$. Numbers in k are designated by Roman, the divisors by German, letters.

I. Integers

Axiom 1. *The unit 1 is an integer.*

Axiom 2. *Sum, difference and product of two integers are integers.*

Axiom 3. *Every number may be written as a fraction*

$$a/b \quad (b \neq 0;\ a,\ b \text{ integers}).$$

II. Divisibility

Axiom 1. *If* $a : \mathfrak{m}$ *and* l *is an integer, then* $la : \mathfrak{m}$. *If* $a : \mathfrak{m}$ *and* $b : \mathfrak{m}$, *then* $(a \pm b) : \mathfrak{m}$.

Definition. $\mathfrak{m} : \mathfrak{b}$ *means that every integer* a *divisible by* $\mathfrak{m}$ *is divisible by* $\mathfrak{b}$.

Axiom 2. *If* $\mathfrak{m} : \mathfrak{b}$ *and* $\mathfrak{b} : \mathfrak{m}$, *then* $\mathfrak{m} = \mathfrak{b}$.

Axiom 1 states that the numbers divisible by a given divisor form an ideal in $[k]$, Axiom 2 that the divisor is uniquely characterized by this corresponding ideal.

Axiom 3. *There exists a divisor* $\mathfrak{n}$ *such that* $1 : \mathfrak{n}$.

Theorem A. *Every integer is divisible by* $\mathfrak{n}$, *and hence* $\mathfrak{n}$ *is unique. (It is called the unit divisor.)*

III. Multiplication of divisors

Axiom 1. $\mathfrak{m}\mathfrak{n} = \mathfrak{m}$.

Axiom 2. $\mathfrak{m}\mathfrak{b} = \mathfrak{b}\mathfrak{m}$.

Axiom 3. $(\mathfrak{m}\mathfrak{b})\mathfrak{c} = \mathfrak{m}(\mathfrak{b}\mathfrak{c})$.

Axiom 4. $(\mathfrak{m}\mathfrak{b}) : \mathfrak{m}$.

Theorem B. $\mathfrak{n} : \mathfrak{m}$ *implies* $\mathfrak{m} = \mathfrak{n}$.

Proof. Combine $\mathfrak{n} : \mathfrak{m}$ with $\mathfrak{m} : \mathfrak{n}$ which follows from Axioms III, 1 and 4.

Axiom 5. $\mathfrak{m} : \mathfrak{b}$ *implies* $\mathfrak{m}\mathfrak{c} : \mathfrak{b}\mathfrak{c}$.

IV. Multiplication of divisor by number

Axiom 1. a *being an integer* $\neq 0$ *and* $\mathfrak{b}$ *a divisor, there exists a divisor* $a\mathfrak{b}$ *such that* $ab : a\mathfrak{b}$ *if and only if* b *is an integer divisible by* $\mathfrak{b}$.

Axiom 2. *Let* a *be an integer* $\neq 0$ *and* $\mathfrak{m}$ *a divisor such that every integer* $: \mathfrak{m}$ *is* $: a$; *then there exists a divisor* $\mathfrak{b}$ *such that* $\mathfrak{m} = a\mathfrak{b}$.

Axiom 3. $a(\mathfrak{m}\mathfrak{b}) = (a\mathfrak{m})\mathfrak{b}$.

Theorem C. $a : \mathfrak{m}$ *is equivalent to* $a\mathfrak{n} : \mathfrak{m}$.

Proof. Axiom II, 1 shows $a : \mathfrak{m}$ to imply $a\mathfrak{n} : \mathfrak{m}$. The converse is trivial because $1 : \mathfrak{n}$.

Theorem D. $\mathfrak{m} : a\mathfrak{n}$ *is equivalent to the statement that every number* $: \mathfrak{m}$ *is* $: a$.

Proof: trivial.

Definition. $a\mathfrak{n}$ *is called the principal ideal* (a).

Theorem E. $a(b\mathfrak{c}) = (ab)\mathfrak{c}$.

Proof. The ideal corresponding to either side consists of all numbers of form $(ab)l$, $l : \mathfrak{c}$.

V. The decisive axioms

The previous axioms are more or less trivial. We now get down to brass tacks.

Axiom 1. *$\mathfrak{a}$ being given, there exist an $\mathfrak{a}'$ such that $\mathfrak{a}\mathfrak{a}'$ is principal.*

Indeed, divisors are intended to serve as ideal factors of numbers. Were Axiom 1 not fulfilled we could, without losing out on the other axioms, limit ourselves to those divisors that show up as factors in principal divisors.

Theorem F. *$\mathfrak{a}\mathfrak{c} : \mathfrak{b}\mathfrak{c}$ implies $\mathfrak{a} : \mathfrak{b}$.*

Proof. $\mathfrak{c}\mathfrak{c}' = (c)$. $c\mathfrak{a} : c\mathfrak{b}$, hence $\mathfrak{a} : \mathfrak{b}$.

Corollary. *$\mathfrak{a}\mathfrak{c} = \mathfrak{b}\mathfrak{c}$ implies $\mathfrak{a} = \mathfrak{b}$.*

Theorem G. *If $\mathfrak{a} : \mathfrak{b}$, then there exists a $\mathfrak{c}$ such that $\mathfrak{a} = \mathfrak{b}\mathfrak{c}$.*

Proof. $\mathfrak{b}\mathfrak{b}' = (b)$. $\mathfrak{a}\mathfrak{b}' : (b)$, hence

$$\mathfrak{a}\mathfrak{b}' = b\mathfrak{c} = (\mathfrak{b}\mathfrak{c})\mathfrak{b}', \qquad \mathfrak{a} = \mathfrak{b}\mathfrak{c}.$$

Theorem H. *$a : \mathfrak{a}$. $b : \mathfrak{b}$ implies $ab : \mathfrak{a}\mathfrak{b}$.*

Proof. $(a) = \mathfrak{a}\mathfrak{a}'$, $(b) = \mathfrak{b}\mathfrak{b}'$, $(ab) = (\mathfrak{a}\mathfrak{b})(\mathfrak{a}'\mathfrak{b}')$.

Definitions. *$\mathfrak{a} = \mathfrak{a}_1 \ldots \mathfrak{a}_h$ is a proper factorization of $\mathfrak{a}$ if no one of the factors $\mathfrak{a}_1$ equals $\mathfrak{n}$. $\mathfrak{p}$ is said to be a prime divisor, if $\mathfrak{p} = \mathfrak{p}$ is its only proper factorization.*

The last two axioms contain the law of unique factorization.

Axiom 2 *(Axiom of limited factorization, frequently formulated as "chain condition"). Given a divisor $\mathfrak{a}$, there exists a natural number h such that $\mathfrak{a}$ allows no proper factorization into more than h factors.*

Axiom 3 *(Axiom of prime divisors). If $\mathfrak{p}$ is prime and $\mathfrak{a}\mathfrak{b} : \mathfrak{p}$ then either $\mathfrak{a} : \mathfrak{p}$ or $\mathfrak{b} : \mathfrak{p}$.*

4. Consequences

Theorem II 4, A. *Any divisor $\mathfrak{a}$ can be split into prime factors $\mathfrak{p}_1$,*

$$\mathfrak{a} = \mathfrak{p}_1 \cdots \mathfrak{p}_h. \tag{4.1}$$

The prime factors are uniquely determined but for their order.

Proof. By splitting $\mathfrak{a}$ into two factors and keeping up this process as long as there are still factors which are not prime, we must come to an end, according to Axiom V, 2.

If

$$\mathfrak{a} = \mathfrak{q}_1\mathfrak{q}_2 \ldots$$

is another prime factorization of $\mathfrak{a}$ besides (4.1), then owing to Axiom V, 3, one of the factors $\mathfrak{p}_1,\ldots,\mathfrak{p}_h$ must be divisible by $\mathfrak{q}_1$, say $\mathfrak{p}_1 : \mathfrak{q}_1$. As $\mathfrak{p}_1$ is prime, $\mathfrak{p}_1$ must equal $\mathfrak{q}_1$, and by canceling the factor $\mathfrak{p}_1 = \mathfrak{q}_1$ one gets

$$\mathfrak{p}_2 \cdots \mathfrak{p}_h = \mathfrak{q}_2 \cdots,$$

and the argument can be iterated.

Theorem II 4, B. *Several ideals $\mathfrak{a}_1,\ldots,\mathfrak{a}_h$ (in particular, several integers which do not all vanish) have a greatest common divisor (GCD).*

Proof. We can write

$$\mathfrak{a}_1 = \mathfrak{p}^{e_1}\mathfrak{p}'^{e_1'} \ldots,$$
$$\cdot\;\cdot\;\cdot\;\cdot\;\cdot\;\cdot\;\cdot\;\cdot\;\cdot$$
$$\mathfrak{a}_h = \mathfrak{p}^{e_h}\mathfrak{p}'^{e_h'}$$

with a finite number of distinct prime factors $\mathfrak{p}, \mathfrak{p}', \ldots$ and exponents $e \geqq 0$. Set

$$e = \min(e_1,\ldots, e_h), \qquad e' = \min(e_1',\ldots, e_h'),\ldots .$$

Then

$$\mathfrak{a} = \mathfrak{p}^{e}\mathfrak{p}'^{e'} \ldots$$

will be a common divisor of $\mathfrak{a}_1,\ldots, \mathfrak{a}_h$, and any common divisor will be a divisor of $\mathfrak{a}$. Notation: $\mathfrak{a} = (\mathfrak{a}_1,\ldots, \mathfrak{a}_h)$.

Theorem II 4, C. *Any divisor $\mathfrak{a}$ is GCD of a finite set of numbers.*

Proof. Choose (Axiom V, 1) $a_1 \neq 0$ divisible by $\mathfrak{a}$ and set $\mathfrak{a}_1 = a_1\mathfrak{o}$. Then $\mathfrak{a}_1 : \mathfrak{a}$. If $\mathfrak{a}_1$ and $\mathfrak{a}$ are not identical, then there exists a number a_2 divisible by $\mathfrak{a}$, but not divisible by $\mathfrak{a}_1$. Introduce $\mathfrak{a}_2 = (\mathfrak{a}_1, a_2)$. This process may be continued, and gives rise to a chain $\mathfrak{a}_1 : \mathfrak{a}_2 : \ldots (: \mathfrak{a})$ and a proper factorization

$$\mathfrak{a}_1 = \frac{\mathfrak{a}_1}{\mathfrak{a}_2} \cdot \frac{\mathfrak{a}_2}{\mathfrak{a}_3} \cdots$$

of $\mathfrak{a}_1$. Hence it must come to a stop after a finite number h of steps with $\mathfrak{a}_h = \mathfrak{a}$.

We now turn to Kronecker's theory which is characterized by a systematic use of indeterminates. We investigate polynomials $f(x,y,..)$ of a given number ν of indeterminates whose coefficients are integers in k ("ganzzahlige" or integral polynomials). Throughout the following, polynomials are supposed to be of this nature. The GCD of the coefficients of $f \neq 0$ is called the <u>content</u> of f and denoted by Ct(f). We have the following significant fact which under more special circumstances is known as Gauss' lemma.

<u>Theorem II 4, D.</u> $Ct(fg) = Ct\, f \cdot Ct\, g$.

<u>Proof</u>. We write

$$\mathfrak{a} = \mathrm{Ct}\, f, \quad \mathfrak{b} = \mathrm{Ct}\, g, \quad \mathfrak{c} = \mathrm{Ct}(fg).$$

Let $\mathfrak{p}$ be any prime divisor and $\mathfrak{a}$ be exactly divisible by $\mathfrak{p}^a$ (i.e., by $\mathfrak{p}^a$ but not by $\mathfrak{p}^{a+1}$) and $\mathfrak{b}$ by $\mathfrak{p}^b$. We will then show $\mathfrak{c}$ to be exactly divisible by $\mathfrak{p}^{a+b}$; that is to say:

(S) If f and g are exactly divisible by $\mathfrak{p}^a$ and $\mathfrak{p}^b$ respectively, then fg is exactly divisible by $\mathfrak{p}^{a+b}$.

Indeed, we order f and g by decreasing powers of x:

$$f = f_0 x^l + f_1 x^{l-1} + \dots ,$$

$$g = g_0 x^m + g_1 x^{m-1} + \dots .$$

Not all the coefficients f_i will be divisible by $\mathfrak{p}^{a+1}$; let f_r be the first coefficient in f not divisible by $\mathfrak{p}^{a+1}$, and g_s the first coefficient in g not divisible by $\mathfrak{p}^{b+1}$. The coefficient $(fg)_{r+s}$ of x^{r+s} in fg will be a polynomial

$$\equiv f_r g_s \pmod{\mathfrak{p}^{a+b+1}}.$$

When the statement (S) holds good for polynomials of one indeterminate less, we can apply it to the two polynomials f_r and g_s of y, ... and find $f_r g_s$ and hence $(fg)_{r+s}$ to be exactly divisible by $\mathfrak{p}^{a+b}$. Thus (S) is proved by induction with respect to the number of indeterminates.

In order to extend the notion of content to rational functions of x, y,..., we introduce fractional divisors. $\frac{\mathfrak{a}}{\mathfrak{b}}$ is simply the pair of the two (integral) divisors $\mathfrak{a}, \mathfrak{b}$ with the convention that

$$\frac{\mathfrak{r}}{\mathfrak{b}} = \frac{\mathfrak{r}'}{\mathfrak{b}'} \quad \text{whenever} \quad \mathfrak{r}\mathfrak{b}' = \mathfrak{r}'\mathfrak{b}.$$

This equality is reflexive, symmetric, and transitive. Multiplication is defined by

$$\frac{\mathfrak{r}}{\mathfrak{b}} \cdot \frac{\mathfrak{r}'}{\mathfrak{b}'} = \frac{\mathfrak{r}\mathfrak{r}'}{\mathfrak{b}\mathfrak{b}'},$$

and its result does not change when one replaces either of the two factors by an equal one. We are justified in identifying the fraction $\frac{\mathfrak{r}}{\mathfrak{n}}$ with the integral divisor $\mathfrak{r}$. The equation $\frac{\mathfrak{r}}{\mathfrak{b}} = \mathfrak{c}$ then means the same as $\mathfrak{r} = \mathfrak{b}\mathfrak{c}$, and $\frac{\mathfrak{r}}{\mathfrak{b}}$ is integral if and only if $\mathfrak{r}$ is divisible by $\mathfrak{b}$.

To any element c of k(xy ..),

$$c = \frac{f(xy\ ..)}{g(xy\ ..)}$$

(f, g polynomials with integral coefficients) we ascribe the content

$$\mathrm{Ct}(c) = \frac{\mathrm{Ct}\ f}{\mathrm{Ct}\ g}\,.$$

This definition is consistent as

$$\frac{f}{g} = \frac{f'}{g'} \qquad \text{or} \qquad fg' = f'g$$

implies

$$\mathrm{Ct}\ f \cdot \mathrm{Ct}\ g' = \mathrm{Ct}\ f' \cdot \mathrm{Ct}\ g \qquad \text{or} \qquad \frac{\mathrm{Ct}\ f}{\mathrm{Ct}\ g} = \frac{\mathrm{Ct}\ f'}{\mathrm{Ct}\ g'},$$

owing to Gauss' generalized lemma, Theorem II 4, D. The content of a product ab is again the product of the contents of the two factors a and b.

An element a of k(xy ..) is said to be integral if its content is an integral divisor, that is to say, if the GCD of the numerator f(xy ..) of a is divisible by the GCD of its denominator. The product of two integers in k(xy ..) is an integer. The same is true for sum and difference. In fact, let $g \neq 0$, $g' \neq 0$ be exactly divisible by $\mathfrak{p}^{v}$, $\mathfrak{p}^{v'}$, and f, f' divisible by $\mathfrak{p}^{u}$, $\mathfrak{p}^{u'}$,

$$(4.2) \qquad u \geqq v, \qquad u' \geqq v',$$

then the numerator of

$$\frac{f}{g} + \frac{f'}{g'} = \frac{fg' + f'g}{gg'}$$

is divisible by the power of $\mathfrak{y}$ with the exponent

$$\min (u + v', u' + v)$$

while the denominator is exactly divisible by the power $v + v'$. The inequalities (4.2) imply

$$u + v' \geqq v + v', \qquad u' + v \geqq v' + v.$$

Thus we have proved

Theorem II 3, E. *The integral elements of $k(x, y \,..)$ form a ring.*

5. Integrity in $\varkappa(xy \,..)$ over $k(xy \,..)$

The notion of integrity always carries with it the notion of unit: a is said to be a unit if a and $\frac{1}{a}$ are integers. Two polynomials $f(xy \,..)$, $g(xy \,..)$ are called associate, $f \sim g$, if $\frac{f}{g}$ is a unit, or what is the same, if $\frac{f}{g}$ and $\frac{g}{f}$ are both integral. $f \sim g$ is the necessary and sufficient condition for f and g to have the same content. We are now in a position to grasp Kronecker's fundamental idea. Instead of operating with divisors, he deals with the polynomials whose contents the divisors are, from the viewpoint that associate polynomials will ultimately be considered equal. Though this aim is constantly before his mind and ours, the identification may be postponed until the final results are reached. On his way he enjoys the greater freedom which polynomials afford, inasmuch as they allow addition and subtraction, besides multiplication. But what is more, we know from §1 how the fundamental notion of integrity carries over from a ground field k to any finite field $\varkappa$ over k. The whole thing goes along a little more smoothly if one replaces integral divisors by arbitrary fractional divisors on the one side, polynomials by rational functions on the other side.

Hence we proceed as follows. We suppose we are given a ground field k and a finite field $\varkappa$ over k; in addition we suppose that integers and divisors are defined in k and satisfy the axioms of §3 so that we may employ all the results of §4 in k but not in $\varkappa$. Our aim is to extend these notions to $\varkappa$ without destroying the validity of the axioms. As an auxiliary construction we use the adjunction of indeterminates. Let k(xy..), $\varkappa$(xy..) be the fields arising from k and $\varkappa$ respectively by adjunction of ν indeterminates x,y, ... The lowest case $\nu = 0$ is not excluded. We generalize to higher ν the fundamental definition of §1:

<u>Definition</u>. *α is said to be an integral element of $\varkappa(xy..)$ if it satisfies an equation*

$$\alpha^r + a_1\alpha^{r-1} + \ldots + a_r = 0$$

whose coefficients are integral elements of $k(xy..)$.

Owing to the fact that the integral elements of k(xy..) form a ring, the same argument as in §1 proves the integral elements of $\varkappa$(xy..) to form a ring. Moreover we can repeat, for elements in $\varkappa$(xy..), the definitions advanced in the previous section:

The integer α is said to be divisible by the integer $\beta \neq 0$ if α/β is integral.

$\alpha \sim \beta$ (α associated with β) means that $\alpha : \beta$ and $\beta : \alpha$.

$\alpha \neq 0$ is a unit if both α and $\frac{1}{\alpha}$ are integers.

We have the obvious fact that

$$\alpha \sim \beta, \quad \alpha' \sim \beta' \quad \text{imply} \quad \alpha\alpha' \sim \beta\beta'$$

(while in general not $\alpha + \alpha' \sim \beta + \beta'$). Our previous suggestions would lead up to this definition:

Any integer α in $\varkappa$(xy..) defines a divisor in $\varkappa$, α and β the same divisor if and only if $\alpha \sim \beta$. Multiplication of divisors is to be effected by means of any of their representing integers in $\varkappa$(xy..).

However, this definition has the disadvantage of depending on the number ν of indeterminates, and there are serious doubts whether it will secure the validity of our axioms; in fact, if we take $\nu = 0$ nothing has been gained at all. We shall, therefore, further modify our process by not fixing the number of indeterminates. But before doing so, let us examine somewhat more closely the integers in $\varkappa$(xy..). Of paramount importance is the following

Theorem II 5, A. *If* α *is an integral element of* $\varkappa(xy..)$, *that is to say, if it satisfies any algebraic equation with integral coefficients in* $k(xy..)$, *then its irreducible equation, and hence its field equation in* $k(xy..)$ *have integral coefficients.*

Proof. Let

$$G(t) = t^h + c_1 t^{h-1} + \dots + c_h$$

be the integral equation for α in $k(xy..)$ and

$$f(t) = a_0 t^g + a_1 t^{g-1} + \dots + a_g$$

the irreducible one; the coefficients a_i of the latter are written as polynomials. $G(t)$ is divisible by $f(t)$ in $k(t, xy..)$. Therefore an equation

$$(5.1) \qquad (a_0 t^g + a_1 t^{g-1} + \dots)(a_0^* t^{h-g} + \dots)$$
$$= b_0 t^h + b_1 t^{h-1} + \dots$$

obtains where the a^* are likewise polynomials and

$$(1 = b_0/b_0), \quad c_1 = b_1/b_0, \quad c_2 = b_2/b_0, \dots$$

are integers. We designate the contents of

$$a_0 \quad \text{and} \quad a_0 t^g + a_1 t^{g-1} + \dots$$

by $\mathfrak{a}_0$ and $\mathfrak{a}$ respectively. Similarly for the second factor and the right side of (5.1). By Theorem II 4, D we have

$$\mathfrak{a}_0 \mathfrak{a}_0^* = \mathfrak{b}_0, \qquad \mathfrak{a}\mathfrak{a}^* = \mathfrak{b}.$$

Obviously $\mathfrak{a}_0 : \mathfrak{a}$, $\mathfrak{a}_0^* : \mathfrak{a}^*$, and our assumption is that not only $\mathfrak{b}_0 : \mathfrak{b}$ but also $\mathfrak{b} : \mathfrak{b}_0$, or $\mathfrak{b} = \mathfrak{b}_0$. Thus one gets

$$\frac{\mathfrak{a}_0}{\mathfrak{a}} \cdot \frac{\mathfrak{a}_0^*}{\mathfrak{a}^*} = \mathfrak{e}.$$

Consequently

$$\frac{\mathfrak{a}_0}{\mathfrak{a}} = \mathfrak{e}, \qquad \mathfrak{a} : \mathfrak{a}_0,$$

or

$$a_0/a_0, \quad a_1/a_0, \ldots, \quad a_g/a_0$$

are integers.

The case of an irreducible equation of degree 1 deserves special emphasis:

Theorem II 5, B. *An element of $k(xy..)$ which is integral in $\varkappa(xy..)$ is an integral element of $k(xy..)$.*

All this holds in particular for numbers ($\nu = 0$), and the last remark settles in the affirmative a question raised at the end of §1.

Our general theorem permits giving the criterion for integrity in $\varkappa(xy..)$ the following neat form: α is an integral element of $\varkappa(xy..)$ if and only if

$$\text{Nm}(t - \alpha) \tag{5.2}$$

is an integral element of $k(t, xy..)$.

Trace and norm of an integer in $\varkappa(xy..)$ are integers in $k(xy..)$. We call

$$\mathfrak{n}(\alpha) = \text{Ct Nm}(\alpha)$$

the divisor norm of α. One has

$$\mathfrak{n}(\alpha\beta) = \mathfrak{n}(\alpha) \cdot \mathfrak{n}(\beta). \tag{5.3}$$

If α is a unit, then $\mathfrak{n}(\alpha)$ is the unit divisor in k. Vice versa, if an integer α of $\varkappa(xy..)$ satisfies the relation $\mathfrak{n}(\alpha) = \mathfrak{N}$, then $1/\alpha$ is integral. To prove this, write α as a quotient of two polynomials φ/ψ. Integrity of α means that

$$\text{Ct Nm}(t\psi - \varphi) \quad \text{or} \quad \text{Ct Nm}(\psi - t\varphi) \tag{5.4}$$

is divisible by Ct Nm(ψ). Integrity of $1/\alpha$ means divisibility of (5.4) by Ct Nm(φ). Both conditions coincide, provided

$$\mathfrak{n}(\alpha) = \mathfrak{N} \quad \text{or} \quad \text{Ct Nm}(\varphi) = \text{Ct Nm}(\psi).$$

Consider an element α of $\varkappa(xy..)$ which is holomorph in x, i.e., of the form

(5.5) $$\alpha = \alpha_0 + \alpha_1 x + \alpha_2 x^2 + \ldots + \alpha_l x^l$$

with the α_i lying in $\varkappa(y..)$.

Theorem II 5, C. *An α holomorph with respect to x is integral in $\varkappa(xy..)$ if and only if the coefficients α_i are integral in $\varkappa(y..)$.*

Proof. An element β of $\varkappa(y..)$ which is integral in $\varkappa(y..)$ is also integral in $\varkappa(xy..)$. That much is trivial. Hence α_i and the terms $\alpha_i x^i$ of (5.5) are integers in $\varkappa(xy..)$ and so is their sum (5.5), provided α_i are integers in $\varkappa(y..)$.

The converse is less obvious. We consider the field equation of α; its coefficients a_i are holomorph in x,

$$a_i = a_{i0} + a_{i1}x + \ldots .$$

If α is an integer, they are moreover integral elements of $k(xy..)$, or the $a_{i0}, a_{i1}, \ldots$ are integral elements of $k(y..)$. By putting $x = 0$ one then finds α_0 to be an integral element of $\varkappa(y..)$. α and α_0 both being integers in $\varkappa(xy..)$, the same holds for

$$\alpha - \alpha_0 = \alpha' = \alpha_1 x + \ldots + \alpha_l x^l$$

and finally for

$$\frac{\alpha'}{x} = \alpha_1 + \alpha_2 x + \ldots + \alpha_l x^{l-1}.$$

Indeed, the equation

$$(\alpha')^h + a_1'(\alpha')^{h-1} + \ldots + a_h' = 0$$

gives rise to the equation

$$t^h + \frac{a_1'}{x} t^{h-1} + \ldots + \frac{a_h'}{x^h}$$

for $t = \alpha'/x$, and

$$a_1'/x, \ldots, a_h'/x^h$$

are integers in $k(xy..)$ if $a_1', \ldots, a_h'$ are so.

Iteration of the process proves our statement that integrity of α implies integrity of $\alpha_0, \alpha_1, \ldots, \alpha_l$.

Corollary *(principle of omitting superfluous variables.) An element α of $\varkappa(y..)$ which is integral in $\varkappa(xy..)$ is also integral in $\varkappa(y..)$.*

Proof: Case $l = 0$ of previous theorem.
From this follows:

Theorem II 5, D. *A polynomial $\varphi(xy..)$ with arbitrary coefficients in $\varkappa$ is integral if and only if all its coefficients are integers.*

6. Kronecker's Theory

(We have used Roman and Greek letters for the numbers in k and $\varkappa$. To denote the divisors in k and $\varkappa$ we should have at our disposal a Roman and a Greek German alphabet. Instead, we shall resort to small and capital German letters.)

As mentioned before, a divisor shall be represented as GCD of the coefficients of a polynomial. By far the most natural way of normalizing this polynomial consists in choosing the linear form with the given coefficients. We thus finally adopt the following fundamental definition as starting point for the theory of divisors:

Definition. *Any finite sequence of integers $\alpha_1, \ldots, \alpha_r$ in $\varkappa$ which do not all vanish, determines a divisor*

$$\mathfrak{A} = (\alpha_1, \ldots, \alpha_r). \tag{6.1}$$

An integer α in $\varkappa$ is said to be divisible by $\mathfrak{A}$ if and only if

$$\frac{\alpha}{\alpha_1 x_1 + \ldots + \alpha_r x_r} \tag{6.2}$$

is an integral element of $\varkappa(x_1 \ldots x_r)$.*

Divisors figure only in basic statements of the form "integer α is divisible by divisor $\mathfrak{A}$," and in such statements as are explicitly defined in terms of basic statements. One therefore has to consider two divisors.

$$\mathfrak{A} = (\alpha_1, \ldots, \alpha_r), \qquad \mathfrak{B} = (\beta_1, \ldots, \beta_s) \tag{6.3}$$

*Erratum for second printing. This sentence should be in Italic type instead of Roman.

equal if every integer divisible by $\mathfrak{a}$ is divisible by $\mathfrak{b}$, and vice versa. $\mathfrak{a}$ is said to be divisible by $\mathfrak{b}$, $\mathfrak{a} : \mathfrak{b}$, if every number divisible by $\mathfrak{a}$ is divisible by $\mathfrak{b}$. Hence $\mathfrak{a} = \mathfrak{b}$ if $\mathfrak{a} : \mathfrak{b}$ and $\mathfrak{b} : \mathfrak{a}$.

Our definition may be stated thus: α is divisible by $\mathfrak{a}$ if α is divisible by the linear form $\alpha_1 x_1 + \ldots + \alpha_r x_r$. In this sense the linear form may stand for the divisor $(\alpha_1, \ldots, \alpha_r)$.

Using the criterion (5.2) we can put the criterion for divisibility of an integer α by the divisor $\mathfrak{a} = (\alpha_1, \ldots, \alpha_r)$ into this neat form:

(6.4)
$$\mathrm{Ct}\ \mathrm{Nm}(\alpha x + \alpha_1 x_1 + \ldots + \alpha_r x_r) : \mathrm{Ct}\ \mathrm{Nm}(\alpha_1 x_1 + \ldots + \alpha_r x_r).$$

Whether α is divisible by $\mathfrak{a}$ can therefore be decided, provided we are able to decide in the ground field k whether the GCD of certain numbers $(a_1, a_2, \ldots)$ is divisible by the GCD of certain other numbers $(b_1, b_2, \ldots)$.

Theorem II 6, A. *Each of the numbers* $\alpha_1, \ldots, \alpha_r$ *is divisible by* $\mathfrak{a} = (\alpha_1, \ldots, \alpha_r)$.

Proof. We show that

(6.5)
$$\frac{\alpha_1 y_1 + \ldots + \alpha_r y_r}{\alpha_1 x_1 + \ldots + \alpha_r x_r}$$

for $y_1 = 1$, $y_2 = \ldots = y_r = 0$ is an integer in $\kappa(x_1 \ldots x_r)$. With the homogeneous form

$$f(x_1 \ldots x_r) = \mathrm{Nm}(\alpha_1 x_1 + \ldots + \alpha_r x_r)$$

in k of degree n one obtains as the field equation of (6.5):

$$\frac{f(tx_1 - y_1, \ldots, tx_r - y_r)}{f(x_1, \ldots, x_r)}.$$

It is evident that the GCD of the numerator is divisible by that of the denominator.

Theorem II 6, B. $\mathfrak{a}$ *is divisible by* $\mathfrak{b}$, *(6.3), if and only if*

(6.6)
$$\frac{\alpha_1 x_1 + \ldots + \alpha_r x_r}{\beta_1 y_1 + \ldots + \beta_s y_s}$$

is integral in $\kappa(x_1 \ldots x_r, y_1 \ldots y_s)$.

Proof. Suppose $\mathfrak{A}$ is divisible by $\mathfrak{B}$. The relation $\alpha_i : \mathfrak{A}$ then implies $\alpha_i : \mathfrak{B}$. In other words

$$\frac{\alpha_i}{\beta_1 y_1 + \ldots + \beta_s y_s} \tag{6.7}$$

are integers in $\varkappa(y_1 \ldots y_s)$. On multiplying by x_i and adding up, one finds (6.6) to be an integer in $\varkappa(x_1 \ldots x_r, y_1 \ldots y_s)$.

Conversely, let (6.6) be integral in $\varkappa(x_1 \ldots x_r, y_1 \ldots y_s)$ and α divisible by $\mathfrak{A}$, so that (6.2) is integral in $\varkappa(x_1 \ldots x_r)$ and a fortiori in $\varkappa(x_1 \ldots x_r, y_1 \ldots y_s)$. Hence the product

$$\frac{\alpha}{\beta_1 y_1 + \ldots + \beta_s y_s}$$

is integral in $\varkappa(x_1 \ldots x_r, y_1 \ldots y_s)$ and, as one omits the superfluous variables $x_1, \ldots, x_r$, also in $\varkappa(y_1 \ldots y_s)$.

One could have defined $\mathfrak{A} : \mathfrak{B}$ by the criterion thus proved. Then the argument used in the proof of Theorem II 6, A, with the y as indeterminates would have shown $\mathfrak{A} : \mathfrak{A}$; and the fact that $\mathfrak{A} : \mathfrak{B}$, $\mathfrak{B} : \mathfrak{C}$ imply $\mathfrak{A} : \mathfrak{C}$ would follow by the principle of omitting superfluous variables.

We have established this criterion for the divisibility of $\mathfrak{A}$ by $\mathfrak{B}$:

$$\text{Ct Nm}(\alpha_1 x_1 + \ldots + \alpha_r x_r + \beta_1 y_1 + \ldots + \beta_s y_s) : \text{Ct Nm}(\beta_1 y_1 + \ldots + \beta_s y_s). \tag{6.8}$$

Theorem II 6, C. *$\mathfrak{A} = (\alpha_1, \ldots, \alpha_r)$ is the greatest common divisor of $\alpha_1, \ldots, \alpha_r$.*

Proof. Indeed, if $\mathfrak{B} = (\beta_1, \ldots, \beta_s)$ is a common divisor of $\alpha_1, \ldots, \alpha_r$, then (6.7) is integral in $\varkappa(y_1 \ldots y_s)$ and (6.6) in $\varkappa(x_1 \ldots x_r, y_1 \ldots y_s)$, or $\mathfrak{A} : \mathfrak{B}$.

Theorem II 6, D. *Two divisors (6.3) have a GCD, namely*

$$(\mathfrak{A}, \mathfrak{B}) = (\alpha_1, \ldots, \alpha_r, \beta_1, \ldots, \beta_s). \tag{6.9}$$

Proof. The argument in Theorem II 6, A, proves

$$\frac{\alpha_1 x_1' + \ldots + \alpha_r x_r' + \beta_1 y_1' + \ldots + \beta_s y_s'}{\alpha_1 x_1 + \ldots + \alpha_r x_r + \beta_1 y_1 + \ldots + \beta_s y_s}$$

to be integral with $y_1' = \dots = y_s' = 0$, or $\mathfrak{A}$ to be divisible by (6.9). Vice versa, if

$$\frac{\alpha_1 x_1 + \dots + \alpha_r x_r}{\gamma_1 z_1 + \dots + \gamma_t z_t}, \qquad \frac{\beta_1 y_1 + \dots + \beta_s y_s}{\gamma_1 z_1 + \dots + \gamma_t z_t}$$

are integral in $\kappa(x,z)$ and $\kappa(y,z)$ respectively, then their sum is integral in $\kappa(x,y,z)$; or with $\mathfrak{A}$ and $\mathfrak{B}$ divisible by $\mathfrak{C}$, (6.9) is divisible by $\mathfrak{C}$.

Multiplication of two linear forms

$$\alpha_1 x_1 + \dots + \alpha_r x_r \quad \text{and} \quad \beta_1 y_1 + \dots + \beta_s y_s$$

yields a bilinear form

$$\Sigma \alpha_i \beta_k x_i y_k \qquad (i = 1, \dots, r;\ k = 1, \dots, s)$$

Hence in order to define the product of two divisors (6.3) it seems necessary to go beyond the bounds of linear forms. Fortunately this is not so because

$$(\Sigma \alpha_i x_i)(\Sigma \beta_k y_k) \sim \Sigma \alpha_i \beta_k u_{ik} \tag{6.10}$$

with rs indeterminates u_{ik}. Indeed, it follows from (6.5) that

$$\frac{\Sigma \alpha_i \beta_k x_i y_k}{\Sigma \alpha_i \beta_k u_{ik}}$$

is integral, and on the other hand, by multiplying

$$\frac{\alpha_i}{\alpha_1 x_1 + \dots + \alpha_r x_r}, \quad \frac{\beta_k}{\beta_1 y_1 + \dots + \beta_s y_s}$$

one realizes that $\alpha_i \beta_k$ and hence $\Sigma \alpha_i \beta_k u_{ik}$ is divisible by the left member of (6.10). We therefore define: An integer γ is divisible by $\mathfrak{A}\mathfrak{B}$ if

$$\frac{\gamma}{\Sigma \alpha_i x_i \cdot \Sigma \beta_k y_k}$$

is integral in $\kappa(x,y)$, and then observe that $\mathfrak{A}\mathfrak{B}$ is the divisor

$$\mathfrak{C} = (\dots, \alpha_i \beta_k, \dots) \quad [i = 1, \dots, r;\ k = 1, \dots, s]$$

in the sense that every integer divisible by $\mathfrak{A}\mathfrak{L}$ is divisible by $\mathfrak{L}$ and vice versa.

One now encounters no difficulty at all in verifying the harmless parts, II to IV, of the axioms in §3. Before settling the decisive axioms V we discuss in what sense the divisors in k are also divisors in $\varkappa$.

Any numbers $a_1, \dots, a_r$ in k give rise to a divisor $(a_1, \dots, a_r) = \mathfrak{A}$ in $\varkappa$ and a divisor $(a_1, \dots, a_r) = \mathfrak{a}$ in k. We indicate this relationship for a moment by $\mathfrak{A} \to \mathfrak{a}$. If the number a in k is divisible by $\mathfrak{A}$ (in $\varkappa$), then it is also divisible by $\mathfrak{a}$ (in k), and vice versa. Moreover, if $\mathfrak{A} \to \mathfrak{a}$, $\mathfrak{L} \to \mathfrak{b}$, then $\mathfrak{A} : \mathfrak{L}$ implies $\mathfrak{a} : \mathfrak{b}$, $\mathfrak{A} = \mathfrak{L}$ implies $\mathfrak{a} = \mathfrak{b}$ (and vice versa),

$$(\mathfrak{A}, \mathfrak{L}) \to (\mathfrak{a}, \mathfrak{b}), \qquad \mathfrak{A}\mathfrak{L} \to \mathfrak{a}\mathfrak{b}.$$

In view of all these facts, a divisor in $\varkappa$ of the form

$$\mathfrak{A} = (a_1, \dots, a_r) \qquad [a_i \text{ in } k]$$

is said to lie in k and is identified with the corresponding divisor $\mathfrak{a}$ in k.

Another relationship between the divisors in k and $\varkappa$ is established by forming the norm: the divisor in k,

$$\mathfrak{a} = \mathfrak{n}(\alpha_1 x_1 + \dots + \alpha_r x_r) = \mathrm{Ct}\ \mathrm{Nm}(\alpha_1 x_1 + \dots + \alpha_r x_r)$$

is said to be the <u>norm</u> of (6.1). Since

$$\mathfrak{n}\left(\frac{\alpha_1 x_1 + \dots + \alpha_r x_r}{\beta_1 y_1 + \dots + \beta_s y_s}\right) = \frac{\mathfrak{n}(\alpha_1 x_1 + \dots + \alpha_r x_r)}{\mathfrak{n}(\beta_1 y_1 + \dots + \beta_s y_s)}$$

$\mathfrak{A} : \mathfrak{L}$ implies $\mathrm{Nm}\,\mathfrak{A} : \mathrm{Nm}\,\mathfrak{L}$, and consequently $\mathfrak{A} = \mathfrak{L}$ implies $\mathrm{Nm}\,\mathfrak{A} = \mathrm{Nm}\,\mathfrak{L}$. The norm is uniquely determined by the divisor, independently of its particular representation as GCD of numbers $\alpha_1, \dots, \alpha_r$. In the same manner there follows from (6.10) the multiplicative law

$$\mathrm{Nm}(\mathfrak{A}\mathfrak{L}) = \mathrm{Nm}\,\mathfrak{A} \cdot \mathrm{Nm}\,\mathfrak{L}.$$

The unit divisor $\mathfrak{E}$ in $\varkappa$ is the only divisor in $\varkappa$ whose norm is the unit divisor $\mathfrak{e}$ in k.

The criterion (6.8) may be stated thus: $\mathfrak{A}$ is divisible by $\mathfrak{L}$ if and only if $\mathrm{Nm}(\mathfrak{A}, \mathfrak{L}) : \mathrm{Nm}\,\mathfrak{L}$.

Let us now proceed to the decisive part V of the axioms!

Axiom 2: *Any factorization of a given divisor* $\mathfrak{a}$ *in* κ,

$$\mathfrak{a} = \mathfrak{a}_1 \dots \mathfrak{a}_h,$$

leads to a corresponding factorization of its norm $\mathfrak{n}$ $= Nm\, \mathfrak{a}$ *into factors* $\mathfrak{n}_1 = Nm\, \mathfrak{a}_1$,

$$\mathfrak{n} = \mathfrak{n}_1 \dots \mathfrak{n}_h,$$

and if $\mathfrak{a}_1 \neq \mathfrak{e}$ *then* $\mathfrak{n}_1 \neq \mathfrak{n}$. *Hence* $\mathfrak{a}$ *in* κ *may not properly be factorized into more factors than its norm* $\mathfrak{n}$ *in* k.

Axiom 3:

$$\mathfrak{a} = (\alpha_1, \dots, \alpha_r), \qquad \mathfrak{b} = (\beta_1, \dots, \beta_s), \qquad \mathfrak{p} = (\pi_1, \dots, \pi_t).$$

If $\mathfrak{p}$ *is prime and* $\mathfrak{a}$ *not divisible by* $\mathfrak{p}$, *then* $(\mathfrak{a}, \mathfrak{p})$ *is the unit divisor or*

$$\frac{1}{\alpha_1 x_1 + \dots + \alpha_r x_r + \pi_1 z_1 + \dots + \pi_t z_t}$$

is integral in $\kappa(x, z)$ *and a fortiori in* $\kappa(x, y, z)$. *Multiply this by*

$$(6.11) \qquad [(\alpha_1 x_1 + \dots + \alpha_r x_r) + (\pi_1 z_1 + \dots + \pi_t z_t)]$$
$$(\beta_1 y_1 + \dots + \beta_s y_s).$$

Under the assumption $\mathfrak{a}\mathfrak{b} : \mathfrak{p}$, *(6.11) and hence the product will be divisible by* $\pi_1 z_1 + \dots + \pi_t z_t$, *or*

$$\frac{\beta_1 y_1 + \dots + \beta_s y_s}{\pi_1 z_1 + \dots + \pi_t z_t}$$

will be an integer in $\kappa(x, y, z)$ *and thus in* $\kappa(y, z)$. *Result: under the hypotheses*

$$\mathfrak{p} \text{ prime}, \ \mathfrak{a}\mathfrak{b} : \mathfrak{p}, \ \mathfrak{a} \text{ not divisible by } \mathfrak{p}$$

the second factor $\mathfrak{b}$ *is divisible by* $\mathfrak{p}$.*

7. The Fundamental Lemma

Up to now we have had smooth sailing and there seems no need at all to transcend the domain of linear forms. It is only for the proof of the one axiom V, 1 that we really have to resort to more general forms. The idea is this. The norm $f(x_1 \ldots x_r)$ of $\alpha_1 x_1 + \ldots + \alpha_r x_r$ splits into head and tail,

$$f(x_1 \ldots x_r) = (\alpha_1 x_1 + \ldots + \alpha_r x_r) \cdot \varphi(x_1 \ldots x_r).$$

The α are supposed to be integers; then the coefficients β_j of φ will be integers in $\varkappa$ while the coefficients γ_l of f will be integers in k whose greatest common divisor $\mathfrak{n}$ is the norm of $\mathfrak{A} = (\alpha_1, \ldots, \alpha_r)$. We introduce the divisor in $\varkappa$,

$$\mathfrak{B} = (\ldots, \beta_j, \ldots)$$

and are going to prove

$$\mathfrak{A}\mathfrak{B} = \mathfrak{n}. \tag{7.1}$$

Axiom V, 1 is assumed to hold in k; hence there exists an $\mathfrak{n}'$ such that $\mathfrak{n}\mathfrak{n}' = (a)$ is principal. $\mathfrak{B}\mathfrak{n}' = \mathfrak{A}'$ in $\varkappa$ then satisfies the equation

$$\mathfrak{A}\mathfrak{A}' = (a),$$

and Axiom V, 1 is proved in $\varkappa$.

The coefficients γ_l of f are combinations of products $\alpha_i\beta_j$ and consequently they are divisible by $\mathfrak{A}\mathfrak{B}$, or $\mathfrak{n} : \mathfrak{A}\mathfrak{B}$.

The difficult part is the converse, $\mathfrak{A}\mathfrak{B} : \mathfrak{n}$ or

$$\frac{\alpha_i\beta_j}{\Sigma\gamma_l z_l} \quad \text{integral.}$$

Its proof depends on the following fundamental

<u>Lemma II 7, A.</u> *Let*

$$\alpha, \alpha_0; \quad \varphi_0, \ldots, \varphi_{h-1}$$

be elements of $\varkappa(x, y, \ldots)$. *If*

$$(\alpha_0 t + \alpha)(\varphi_0 t^{h-1} + \ldots + \varphi_{h-1}) = \psi_0 t^h + \psi_1 t^{h-1} + \ldots + \psi_h \tag{7.2}$$

has integral coefficients ψ, *then the products*

$$\alpha_0\varphi_0, \ldots, \alpha_0\varphi_{h-1}$$

are also integral.

Proof. We exclude the trivial case $\alpha_0 = 0$.

$\alpha_0\varphi_0 = \psi_0$ is integral. $t = -\alpha/\alpha_0$ is a root of the polynomial (7.2); consequently

$$\rho = \psi_0 \cdot \frac{\alpha}{\alpha_0} = \alpha\varphi_0$$

satisfies the following equation with integral coefficients in $\varkappa(xy..)$:

$$\rho^h - \psi_1\rho^{h-1} + \psi_2\psi_0\rho^{h-2} - \ldots \pm \psi_h\psi_0^{h-1} = 0,$$

and thus by Theorem II 1, A is integral. The coefficients of

$$(\alpha_0 t + \alpha)(\varphi_1 t^{h-2} + \ldots + \varphi_{h-1}) = \psi_1' t^{h-1} + \ldots + \psi_h'$$

are

$$\psi_1' = \psi_1 - \alpha\varphi_0, \quad \psi_2' = \psi_2, \ldots, \quad \psi_h' = \psi_h$$

and therefore integral. By induction with respect to the degree h we thus prove our lemma.

We are interested in the case where the coefficients $\psi_0, \ldots, \psi_h$ lie in k. The equations with integral coefficients in k which we obtain for

$$\alpha_0\varphi_0, \ \alpha_0\varphi_1, \ \alpha_0\varphi_2, \ \ldots$$

by following the construction in the proof of Theorem II 1, A will then be of the degrees

$$1, \quad h, \quad h(h - 1), \ \ldots$$

respectively.

With $t = x_1$ and

$$\frac{\alpha_1 x_1 + \ldots + \alpha_r x_r}{\Sigma\gamma_l z_l} \quad \text{and} \quad \varphi(x_1 \ldots x_r)$$

as the first and second factor,

$$\alpha_1/\Sigma\gamma_l z_l = \alpha_0, \quad (\alpha_2 x_2 + \ldots + \alpha_r x_r)/\Sigma\gamma_l z_l = \alpha,$$

we apply the lemma and find that after ordering φ according to powers of x_1,

$$\varphi(x_1 \ldots x_r) = \varphi_0 x_1^{n-1} + \ldots + \varphi_{n-1},$$

the products

$$\alpha_0\varphi_0, \ldots, \alpha_0\varphi_{n-1}$$

are integral. These elements of $\varkappa(z;\ x_2 \ldots x_r)$ are holomorph in $x_2, \ldots, x_r$ and thus by Theorem II 5, D, all its coefficients $\alpha_0\beta_j$ or

$$\alpha_1\beta_j/\Sigma\gamma_l z_l$$

are integral in $\varkappa(z)$. α_1 may here be replaced by any of the other coefficients $\alpha_2, \ldots, \alpha_r$.

After having settled the last axiom we can carry over to $\varkappa$ Gauss' generalized lemma, Theorem II 4, D, of which the previous lemma is a particular case.

It should be kept in mind that Nm $\mathfrak{A}$ is divisible by $\mathfrak{A}$ in $\varkappa$.

We are now able to prove quite generally:

Theorem II 7, B. *If $\alpha_1, \ldots, \alpha_r$ are the coefficients of a polynomial $\varphi(x, y, \ldots)$ in $\varkappa$, then*

(7.3) $$\varphi(x, y, \ldots) \sim \alpha_1 u_1 + \ldots + \alpha_r u_r.$$

This is the true reason why the restriction to linear forms has been successful. The integrity of

$$\frac{\varphi(x, y, \ldots)}{\alpha_1 u_1 + \ldots + \alpha_r u_r}$$

follows from Theorem II 6, A. It is far less evident that

$$\sigma = \frac{\alpha_1 u_1 + \ldots + \alpha_r u_r}{\varphi(x, y, \ldots)}$$

is integral in $\varkappa(u_1 \ldots u_r,\ xy \ldots)$. Consider the norm of φ,

$$f(x, y, \ldots) = \text{Nm}\ \varphi = \varphi(x, y, \ldots) \cdot T_\varphi(x, y, \ldots).$$

Let α_i, β_j, c_l denote the coefficients of φ, $\tau = T_\varphi$, and f respectively, and $\mathfrak{A}$, $\mathfrak{B}$, $\mathfrak{c}$ the contents of these polynomials. Then we know that

$$\mathfrak{A}\mathfrak{B} = \mathfrak{c},$$

in particular $\mathfrak{A}\mathfrak{B} : \mathfrak{c}$, that is to say the products $\alpha_i\beta_j$ are divisible by $\mathfrak{c}$, or

$$\alpha_i\beta_j/\Sigma c_l v_l \quad \text{are integers in } \varkappa(v).$$

Therefore

$$\rho = \frac{(\alpha_1 u_1 + \ldots + \alpha_r u_r) \cdot \tau(xy\ldots)}{\Sigma c_l v_l}$$

is integral in $\varkappa(u, v; xy\ldots)$. We wish to prove the integrity of

$$\sigma = \frac{(\alpha_1 u_1 + \ldots + \alpha_r u_r) \cdot \tau(xy\ldots)}{f(xy\ldots)} = \rho b \qquad \text{with}$$

$$b = (\Sigma c_l v_l)/f(xy\ldots).$$

By definition, b is integral in $k(v; xy\ldots)$. The equation for ρ,

$$\rho^h + a_1\rho^{h-1} + \ldots + a_h = 0 \quad [a_i \text{ integers in } k(u, v; xy\ldots)]$$

implies the equation

$$\sigma^h + a_1 b \cdot \sigma^{h-1} + \ldots + a_h b^h = 0$$

for σ whose coefficients are likewise integers. Hence σ is integral first in $\varkappa(u, v; xy\ldots)$ and then in $\varkappa(u; xy\ldots)$.

According to (7.3) the quotient

$$\frac{\varphi(x, y, \ldots)}{\psi(x, y, \ldots)}$$

of two polynomials φ, ψ with integral coefficients α_i, β_j in $\varkappa$ and with the contents $\mathfrak{A}$, $\mathfrak{B}$ is integral if and only if

$$(\Sigma\alpha_i u_i)/(\Sigma\beta_j v_j)$$

is integral, i.e., if $\mathfrak{A} : \mathfrak{B}$. We thus fall back in $\varkappa$ upon that criterion of integrity which served as a definition in k, and the cycle of our argumentation is complete.

The result is this. If the notions of integer and divisor are established in the ground field k in such a way that besides obvious requirements they satisfy the law of unique factorization into primes, then it is possible to extend these notions in the same manner to any finite field $\varkappa$ over k; the word "extend" indicating that the integers and divisors of k are among those of $\varkappa$ and the relations established between them by addition, multiplication and divisibility are not affected by passing from k to $\varkappa$.

8. A Batch of Simple Propositions

I. Consider a field k in which the axioms of §3 hold, and let $\mathfrak{p}$ be a prime divisor in k. The residues mod $\mathfrak{p}$ then form a ring $k_{\mathfrak{p}}$ without null divisors; it consists of all integers of k with the convention that two integers a, b are considered equal if

$$a \equiv b \pmod{\mathfrak{p}}$$

or a - b is divisible by $\mathfrak{p}$.

By applying to this field a well-known theorem about roots of a polynomial in a ring without null divisors we find

Theorem II 8, A. *A congruence of degree m,*

$$f(x) \equiv 0(\mathfrak{p}); \quad f(x) = a_0x^m + \dots + a_m, \quad a_0 \not\equiv 0(\mathfrak{p}),$$

has at most m incongruent solutions in k. If it has the l incongruent solutions $c_1, \dots, c_l$ then

$$(8.1) \qquad f(x) \equiv (x - c_1) \dots (x - c_l) \cdot g(x) \pmod{\mathfrak{p}}$$

where g(x) is of degree m - l.

A congruence like (8.1) for polynomials is, of course, to be interpreted as stating that left and right members are identical in the field $k_{\mathfrak{p}}$, and identity of polynomials means that corresponding coefficients coincide.

Theorem II 8, B. *Given the divisors $\mathfrak{a}$, $\mathfrak{b}$ there exists an integer a divisible by $\mathfrak{a}$ such that $(a)/\mathfrak{a}$ is prime to $\mathfrak{b}$.*

Proof. Suppose we have the decomposition into powers of distinct prime divisors:

$$\mathfrak{a} = \mathfrak{p}_1^{e_1}\mathfrak{p}_2^{e_2} \dots,$$
$$\mathfrak{b} = \mathfrak{p}_1^{d_1}\mathfrak{p}_2^{d_2} \dots .$$

Select an integer a_1 divisible by

$$\mathfrak{p}_1^{e_1}\mathfrak{p}_2^{e_2+1}\mathfrak{p}_3^{e_3+1} \dots$$

but not divisible by

$$\mathfrak{p}_1^{e_1+1}\mathfrak{p}_2^{e_2+1}\mathfrak{p}_3^{e_3+1} \dots .$$

The integer

$$a = a_1 + a_2 + \dots$$

will do the trick. It is divisible by $\mathfrak{r}$, and all summands except the first are even divisible by $\mathfrak{p}_1\mathfrak{r}$. But the first is certainly not. Indeed

$$\frac{(a_1)}{\mathfrak{r}} = \frac{(a_1)}{\mathfrak{p}_1^{e_1}\mathfrak{p}_2^{e_2+1}\mathfrak{p}_3^{e_3+1}\dots} \cdot \mathfrak{p}_2\mathfrak{p}_3 \dots$$

is indivisible by $\mathfrak{p}_1$ because the first factor of the right member and all the following factors $\mathfrak{p}_2, \mathfrak{p}_3, \dots$ are indivisible by $\mathfrak{p}_1$. Hence $\frac{(a)}{\mathfrak{r}}$ is indivisible by $\mathfrak{p}_1$, and for similar reasons by $\mathfrak{p}_2, \dots$. Therefore it has no common prime divisor with $\mathfrak{b}$.

Theorem II 8, C. *If the integer a is divisible by the divisor $\mathfrak{r}$, then one can ascertain another integer b such that $\mathfrak{r} = (a, b)$.*

Proof. Choose b divisible by $\mathfrak{r}$ such that $\frac{(b)}{\mathfrak{r}}$ is prime to $\frac{(a)}{\mathfrak{r}}$.

This looks like an enormous simplification if compared with Theorem II 4, C, but as a matter of fact it has little practical effect.

II. We next pass to a finite field $\varkappa$ over k of degree n.

Theorem II 8, D. *The norm of a prime divisor $\mathfrak{P}$ in $\varkappa$ is a power $\mathfrak{p}^f$ of a prime divisor $\mathfrak{p}$ in k. The exponent f is called the degree of $\mathfrak{P}$ (relative to k).*

Proof. Nm $\mathfrak{P} = \mathfrak{r}$ is divisible by $\mathfrak{P}$. Decompose $\mathfrak{r}$ in k into its prime factors $\mathfrak{r} = \mathfrak{p}\mathfrak{q} \dots$. One of them at least, say $\mathfrak{p}$, must be divisible by $\mathfrak{P}$, $\mathfrak{p} = \mathfrak{P}\mathfrak{L}$. In taking the norm of this equation one gets

$$\mathfrak{p}^n = \mathrm{Nm}\,\mathfrak{P} \cdot \mathrm{Nm}\,\mathfrak{L} .$$

But the only factors of $\mathfrak{p}^n$ in k are powers of $\mathfrak{p}$.

Theorem II 8, E. *Let*

(8.2) $$\mathfrak{p} = \mathfrak{P}_1^{e_1}\mathfrak{P}_2^{e_2}\ \ldots$$

be the decomposition of the prime divisor $\mathfrak{p}$ *in* k *into powers of distinct prime divisors* $\mathfrak{P}_1, \mathfrak{P}_2, \ldots$ *in* κ *of the respective degrees* $f_1, f_2, \ldots$. *Then*

(8.3) $$n = e_1 f_1 + e_2 f_2 + \ldots .$$

Proof by taking the norm of the equation (8.2).

Corollary: *The number* $e_1 + e_2 + \ldots$ *of prime factors of* $\mathfrak{p}$ *in* κ *cannot exceed* n.

Here is an important application of the last proposition to the theory of cyclotomic fields. Let l be a (positive) rational prime number. We study the equation

$$x^l - 1 = 0$$

in the field $\mathfrak{g}$ of common rational numbers. Its roots in the Gaussian complex plane are the vertices of the regular l-gon,

$$e\left(\frac{g}{l}\right) \quad [g = 0, 1, \ldots, l - 1], \qquad e(x) = e^{2\pi i x}.$$

In a purely algebraic treatment we first split off the root 1:

(8.4) $$(x - 1)(x^{l-1} + \ldots + x + 1).$$

Let $k = \mathfrak{g}(\zeta)$ be the field defined by an irreducible factor $f(x)$ of

(8.5) $$x^{l-1} + \ldots + x + 1, \qquad f(\zeta) = 0.$$

$\zeta^l = 1$, but no lower power of ζ equals 1. Indeed if ζ^m is the lowest power = 1, the only powers of that sort are those with exponents divisible by m. Hence m is a divisor of l, and as 1 is excluded, m = l. Consequently

$$\zeta, \zeta^2, \ldots, \zeta^{l-1}$$

are distinct one from the other and from 1. Being roots of $x^l - 1$, they must be roots of the second factor in (8.4), hence the decomposition in k:

$$x^{l-1} + \ldots + x + 1 = (x - \zeta)(x - \zeta^2) \ldots (x - \zeta^{l-1}).$$

Substitute x = 1:

$$(8.6) \qquad l = (1 - \zeta)(1 - \zeta^2) \ldots (1 - \zeta^{l-1}).$$

We put

$$1 - \zeta = \lambda \text{ and } \mathfrak{l} = (\lambda).$$

$$\frac{1 - \zeta^g}{1 - \zeta} \quad (g = 1, \ldots, l - 1)$$

is a unit in k. Indeed, it is integral because it equals

$$1 + \zeta + \ldots + \zeta^{g-1}.$$

On the other side

$$\frac{1 - \zeta}{1 - \zeta^g} = \frac{1 - \zeta'^{g'}}{1 - \zeta'} = 1 + \zeta' + \ldots + \zeta'^{g'-1}$$

with $\zeta^g = \zeta'$ and $gg' \equiv 1 \pmod{l}$. Hence (8.6) implies

$$(l) = \mathfrak{l}^{l-1}.$$

In view of our last theorem, this shows:

Theorem II 8, F. *The degree of the cyclotomic field $\mathfrak{I}(\zeta)$ is $l - 1$, or the polynomial (8.5) is irreducible in $\mathfrak{I}$. $\mathfrak{l}$ is a prime ideal of degree 1 in $\mathfrak{I}(\zeta)$.*

We return to the general theory. The following proposition is sometimes useful for ascertaining the degree f of a prime divisor $\mathfrak{P}$ in a field $\varkappa$ over k.

Theorem II 8, G. *The norm of every integer divisible by $\mathfrak{P}$ is divisible by $\mathfrak{p}^f = Nm\ \mathfrak{P}$, and there are integers $\pi : \mathfrak{P}$ for which $Nm\ \pi$ is divisible by no higher power of $\mathfrak{p}$.*

The first part is obvious. Decompose $\mathfrak{p}$ into its prime factors in $\varkappa$,

$$\mathfrak{p} = \mathfrak{P} \cdot \mathfrak{P}_1 \mathfrak{P}_2 \ldots$$

and choose π divisible by $\mathfrak{P}$ so that $\frac{(\pi)}{\mathfrak{P}}$ is prime to $\mathfrak{p}$:

$$(\pi) = \mathfrak{P}\,\mathfrak{Q}_1\mathfrak{Q}_2\ \ldots, \quad \mathrm{Nm}\ \pi = \mathfrak{p}^f \cdot \mathfrak{q}_1^{f_1}\mathfrak{q}_2^{f_2}\ \ldots\ .$$

As $\mathfrak{Q}_1, \mathfrak{Q}_2, \ldots$ do not go into $\mathfrak{p}$, the prime divisors $\mathfrak{q}_1$, $\mathfrak{q}_2, \ldots$ are different from $\mathfrak{p}$ and hence π satisfies our requirement.

9. Relative Norm of a Divisor

Our notation is very awkward when we deal with a two-storied tower $k \subset \varkappa \subset K$. We shall then denote divisors in k, $\varkappa$, K by $\mathfrak{a}$, $\mathfrak{a}$, $\mathfrak{A}$ respectively. k is considered the ground field.

Theorem II 9, A. $Nm_K(\mathfrak{A}) = Nm_\varkappa(Nm_{K/\varkappa}\ \mathfrak{A})$.

Proof. If $\mathfrak{A} = (A_1, \ldots, A_r)$, then $\mathrm{Nm}_K\mathfrak{A}$ and $\mathrm{Nm}_{K/\varkappa}\mathfrak{A}$ are the contents of the forms

$$\mathrm{Nm}_K(A_1x_1 + \ldots + A_rx_r), \quad \mathrm{Nm}_{K/\varkappa}(A_1x_1 + \ldots + A_rx_r)$$
$$= \varphi(x_1 \ldots x_r)$$

in k and $\varkappa$ respectively. With α_j being the coefficients of φ we have

$$\varphi(x_1 \ldots x_r) \sim \sum_j \alpha_j u_j,$$

and by taking the norm,

$$\mathrm{Nm}_\varkappa(\varphi(x_1 \ldots x_r)) \sim \mathrm{Nm}_\varkappa(\Sigma\alpha_j u_j).$$

The content of the left member

$$\mathrm{Nm}_\varkappa\left\{\mathrm{Nm}_{K/\varkappa}(A_1x_1 + \ldots + A_rx_r)\right\} = \mathrm{Nm}_K(A_1x_1 + \ldots + A_rx_r)$$

is $\mathrm{Nm}_K\mathfrak{A}$ while by definition the content of the right side is $\mathrm{Nm}_\varkappa(\mathrm{Nm}_{K/\varkappa}\mathfrak{A})$.

10. The Dedekind Case

Let us suppose the following statement to be true in the ground field k: Whenever the integers $a_1, \ldots, a_r$ are without common divisor,

$$(a_1, \ldots, a_r) = (1),$$

then there exist integers l_i such that

$$a_1 l_1 + \dots + a_r l_r = 1.$$

Under this assumption k is said to be a Dedekind field In- deed, Dedekind's theory was devised to meet this case. The following propositions assume k to be a Dedekind field.

Theorem II 10, A. *If a is divisible by the divisor $\mathfrak{m} = (a_1, \dots, a_r)$ then there exist integers l_i such that*

$$a = a_1 l_1 + \dots + a_r l_r. \tag{10.1}$$

Proof. Choose a divisor $\mathfrak{b} = (b_1, \dots, b_s)$ such that $\mathfrak{m}\mathfrak{b} = (b)$ is principal. Then $ab_j : b$ and we set $ab_j/b = c_j$. The GCD of all integers

$$\frac{a_i b_j}{b} \quad (i = 1, \dots, r;\ j = 1, \dots, s)$$

is 1. Hence we may ascertain integers l_{ij} such that

$$\sum_{i,j} \frac{a_i b_j}{b} \cdot l_{ij} = 1.$$

After multiplying by a and introducing

$$\sum_j l_{ij} c_j = l_i$$

one gets (10.1).

Theorem II 10, B. *A finite field $\varkappa$ over a Dedekind field k is a Dedekind field.*

Proof. Let $\alpha_1, \dots, \alpha_r$ be integers in $\varkappa$ such that

$$(\alpha_1, \dots, \alpha_r) = (1).$$

$$Nm(\alpha_1 x_1 + \dots + \alpha_r x_r) = f(x_1 \dots x_r)$$

then is a primitive form, that is to say one whose coefficients a_j are without common divisor. The tail φ in

$$f(x_1 \dots x_r) = (\alpha_1 x_1 + \dots + \alpha_r x_r) \cdot \varphi(x_1 \dots x_r)$$

is a form in $\varkappa$ with integral coefficients. Hence each coefficient a_j of f is of the form

$$a_j = \sum_i \lambda_{ji}\alpha_i \qquad (\lambda_{ji} \text{ integers in } \varkappa).$$

By hypothesis there can be found integers l_j in k such that

$$\Sigma a_j l_j = 1.$$

Then

$$\sum_i \alpha_i \lambda_i = 1$$

with

$$\lambda_i = \sum_j l_j \lambda_{ji}.$$

Theorem II 10, C. *The congruence*

$$ax \equiv b \quad (\mathfrak{m})$$

for x is solvable if b : (a, $\mathfrak{m}$).

Proof. $\mathfrak{m} = (a_1, \ldots, a_r)$. The hypothesis implies the existence of numbers $l, l_1, \ldots, l_r$ such that

$$b = al + a_1 l_1 + \ldots + a_r l_r$$

or

$$b \equiv al \quad (\text{mod } \mathfrak{m}).$$

Corollary. *If $\mathfrak{p}$ is prime and $a \not\equiv 0(\mathfrak{p})$ the congruence*

$$ax \equiv b(\mathfrak{p})$$

has a solution x which is uniquely determined mod $\mathfrak{p}$. Hence the residue classes modulo a prime divisor $\mathfrak{p}$ not only form a ring without null divisors but even a field.

Theorem II 10, D. *If the divisors $\mathfrak{m}_1, \ldots, \mathfrak{m}_l$ are relatively prime in pairs, then the simultaneous congruences*

$$(10.2) \quad x \equiv a_1 \quad (\mathfrak{m}_1), \ldots, \; x \equiv a_l \quad (\mathfrak{m}_l)$$

have a solution x which is uniquely determined mod $\mathfrak{m}_1 \cdots \mathfrak{m}_l$.

Proof. We can ascertain an integer b_1 divisible by $\mathfrak{m}_2 \ldots \mathfrak{m}_l$ such that

$$\frac{(b_1)}{\mathfrak{m}_2 \ldots \mathfrak{m}_l} \quad \text{is prime to } \mathfrak{m}_1,$$

and hence b_1 is prime to $\mathfrak{m}_1$. Indeed, were

$$\mathfrak{m}_1 : \mathfrak{p}, \qquad b_1 : \mathfrak{p}, \qquad \mathfrak{p} \text{ prime},$$

one would have

$$\frac{(b_1)}{\mathfrak{m}_2 \ldots \mathfrak{m}_l} \cdot \mathfrak{m}_2 \ldots \mathfrak{m}_l \quad \text{divisible by } \mathfrak{p},$$

and since $\mathfrak{m}_2, \ldots, \mathfrak{m}_l$ are not divisible by $\mathfrak{p}$, the first factor would be, contrary to construction. We set

$$x = b_1 x_1 + \ldots + b_l x_l,$$

and the conditions (10.2) reduce to

$$b_1 x_1 \equiv a_1 \quad (\mathfrak{m}_1), \ldots, \qquad b_l x_l \equiv a_l \quad (\mathfrak{m}_l).$$

Each of these congruences is solvable, on account of $(b_1, \mathfrak{m}_1) = (1)$.

The uniqueness of x (mod $\mathfrak{m}_1 \ldots \mathfrak{m}_l$) is equivalent to the statement that a number a divisible by $\mathfrak{m}_1$ and $\mathfrak{m}_2$ .. and $\mathfrak{m}_l$ is divisible by their product $\mathfrak{m}_1 \ldots \mathfrak{m}_l$ provided the $\mathfrak{m}$'s are prime in pairs. This fact is readily proved by decomposing the $\mathfrak{m}_i$ into powers of distinct primes; it has nothing to do with the Dedekind assumption and should perhaps better have found its place in §4.

11. Kronecker and Dedekind

More than sixty years ago they were rivals striving for the common goal of founding a general arithmetic of algebraic number fields. They reached it along different roads. To which of the two victors shall we give the palm?

What they had in mind was above all the classical case of number theory with the ground field k = $\mathfrak{Q}$. We may add at once the case of the ground field Ω(X) consisting of all rational functions of a single variable X with arbitrary complex coefficients. (We denote this variable by a capital so as to distinguish it from the auxiliary indeterminates in Kronecker's construction.) In the first case the integers are the common integers 0, ± 1, $\pm 2, \ldots$, in the second case the polynomials of X. The only divisors are the integers themselves, since they satisfy the law of unique prime factorization. Both fields are of Dedekind nature. A finite field $\varkappa$ over k is an "algebraic number field," or an "algebraic function field with one variable X" respectively.

k being a Dedekind field, the criterion, Theorem II, 10, A, holds good in $\varkappa$. Dedekind takes it as his starting point; he defines:

Any finite sequence of integers $\alpha_1, \ldots, \alpha_r$ in $\varkappa$ determines a divisor (or ideal) $\mathfrak{m}$; the integer α is said to be divisible by $\mathfrak{m}$ if there exist integers $\lambda_1, \ldots, \lambda_r$ such that

$$\alpha = \alpha_1\lambda_1 + \ldots + \alpha_r\lambda_r. \tag{11.1}$$

Compare this with Kronecker's definition at the beginning of §6. There is a glaring coincidence and a glaring difference: for Kronecker's indeterminates $x_1, \ldots, x_r$ Dedekind substitutes arbitrary integers $\lambda_1, \ldots, \lambda_r$; the divisibility of α by $\alpha_1 x_1 + \ldots + \alpha_r x_r$ is replaced by the equation (11.1).

As both theories are actually equivalent one can dissent about questions of convenience only. To my judgment the odds are here definitely against Dedekind. His theory suffers from a certain lack of self-sufficiency, in so far as its proofs resort to indeterminates and pivot around the fundamental Lemma II 7, A, tools which are native to Kronecker's set up, alien to Dedekind's. A proof of Theorem II 8, A, so simple by means of forms and their contents, seems nearly impossible without this instrument. Kronecker's criterion of divisibility is one decidable by finite means, while Dedekind's criterion refers to the infinite set of all possible integers λ. This has further awkward consequences. The question whether the number α in $\varkappa$ is divisible by the divisor $(\alpha_1, \ldots, \alpha_r)$ in $\varkappa$ is answered by Dedekind in different ways according to whether the question is put in $\varkappa$ or in a finite field K over $\varkappa$, the answer requiring solvability of the equation

$$\alpha = \alpha_1\Lambda_1 + \ldots + \alpha_r\Lambda_r$$

by integers Λ_i in K in the second, by integers Λ_i in $\varkappa$ in the first case. It is a remote consequence of the theory that both requirements agree, while in Kronecker's theory the embedding field, $\varkappa$ or K, is irrelevant for the definition. (In a joint paper with H. Weber, laying the foundations for an arithmetical theory of the algebraic functions of one variable, Dedekind himself adopted a method closely related to Kronecker's approach; cf. H. Weber, Lehrbuch der Algebra, 2d ed., Braunschweig 1908, vol. 3, 5th book.)

The situation becomes much more serious if the two competing theories are stretched to cover cases of non-Dedekind ground fields. The most important is that one with which the algebraic geometers are concerned:

elements of $k = \Omega(X_1 \ldots X_m)$ are the rational functions of $m > 1$

variables $X_1, \ldots, X_m$ with arbitrary complex coefficients,

integers = divisors of k are the polynomials of $X_1, \ldots, X_m$ which form the ring $\Omega[X_1 \ldots X_m]$.

Let us take for an example $m = 3$. A single polynomial equation

$$f(X_1X_2X_3) = 0$$

defines an algebraic surface in the 3-space E_3 with the coördinates X_1, X_2, X_3. The law of unique decomposition states that any such algebraic surface splits in a unique manner into irreducible surfaces. By passing from k to a field $\varkappa$ over k of finite degree n we pass from the space E_3 to an algebraic 3-space F_3 (of degree n over E_3). Kronecker's theory tells how to define algebraic surfaces in F_3 in such a manner that the law of unique decomposition into irreducible constituents still prevails: the divisors are to represent the surfaces; $\mathfrak{a} : \mathfrak{b}$ is interpreted in geometric language as "surface $\mathfrak{b}$ is part of $\mathfrak{a}$," the product $\mathfrak{a}\mathfrak{b}$ as the "union" of the two surfaces $\mathfrak{a}$ and $\mathfrak{b}$. Thus Kronecker's theory amounts to a very reasonable geometry of surfaces in F_3.

Any ideal in the ring $[k] = \Omega[X_1X_2X_3]$ represents an "algebraic manifold" in E_3. The geometer would attempt to describe it as a thing consisting of surfaces, curves and points. Since the law of unique decomposition does not hold for these ideals, the word "consisting" in the previous sentence remains somewhat vague. The ideals in the ring of integers of a $\varkappa$ represent algebraic manifolds in F_3. When we want a clearer distinction we oppose these D-manifolds defined by <u>D</u>edekind ideals, to the K-surfaces defined by Kronecker divisors. For ideals we also have a notion of divisibility and multiplication, and hence the notions of "D-part" and "D-union" for D-manifolds. The D theory of manifolds and the K theory of surfaces move on two entirely different levels. The issue has been hopelessly confounded by the geometric language which is so inadequate for both theories, because it carries the false suggestion that the entities under discussion are point-sets.

In spite of the chasm, there are important bridges between the two realms of divisors and ideals, of K-surfaces and D-manifolds in F_3. Any divisor $\mathfrak{a}$ determines an ideal, namely the ideal of the integers divisible by $\mathfrak{a}$. Hence every K-surface coincides with a D-manifold; in this sense the surfaces are special manifolds. One surface may be part of another, and it does not matter whether we understand "part" in the K or the D sense. Indeed, divisibility $\mathfrak{a} : \mathfrak{b}$ has the same meaning for two divisors $\mathfrak{a}, \mathfrak{b}$ and the corresponding ideals, for it states that every integer divisible by $\mathfrak{a}$ is divisible by $\mathfrak{b}$. But the "greatest common K and D parts" $(\mathfrak{a}, \mathfrak{b})$ of two surfaces in general do not agree. For instance the surfaces $X_1 = 0$ and $X_2 = 0$ have no common part in the K sense [the Kronecker divisor $(X_1, X_2) = (1)$], while in the D sense they have a straight line in common [the Dedekind ideal $(X_1, X_2) \neq (1)$]. There is no reason why the K union of two surfaces in F_3 should always be the same as their D union (although this is true in E_3 where there are no other than principal divisors).

Besides the coincidence of surfaces with special manifolds, there is another interconnection between manifolds and surfaces. I do not know whether every ideal in $\varkappa$ has a finite ideal basis (as in k). Anyhow, from any ideal with finite ideal basis $(\alpha_1, \dots, \alpha_r)$ we may derive the Kronecker divisor $(\alpha_1, \dots, \alpha_r)$. Any integer contained in the ideal is divisible by this corresponding divisor (but by no means vice versa). The relations

$$\mathfrak{a} : \mathfrak{b} \quad \text{and hence} \quad \mathfrak{a} = \mathfrak{b}$$

for ideals are reflected in the same relations for the corresponding divisors. In geometric language: every manifold (with finite ideal basis) is associated with a uniquely determined surface which is part of it. (Even if the manifold itself coincides with a surface, the associated surface may be a proper part of it.)

All this awaits closer investigation. Of the D theory we have here not even given the first beginnings.

The discrepancy between the K and D aspects probably increases if an arbitrary field serves as coefficient field for the polynomials $f(X_1X_2X_3)$ rather than the field Ω. In summarizing, one may venture to say that K is the more fundamental, D the more complete theory; or that D is of higher importance to the geometer, who ought to be concerned about manifolds of every dimension, while K is more important to the arithmetician, whose chief concern (presuming he is old-fashioned enough!) is the law of unique factorization.

We return to the case $m = 1$. E_1 has one complex or two real dimensions; if it is depicted as the Gaussian plane, then F_1 appears as a Riemann surface over E_1 with n sheets. For the coefficient field Ω the only prime polynomials in k are the linear factors $X - a$, and the equation $X - a = 0$ defines the one point a. The 0-dimensional manifolds on F_1 to which the Kronecker and the Dedekind theories here lead in harmonious unison, are finite sets of points on F_1: an integer α in $\varkappa$ is divisible by the divisor $\mathfrak{m}$ or contained in the corresponding ideal if and only if the function α vanishes in all the points of the corresponding group. Very likely its dissolution into individual points will correspond to the decomposition of the divisor into prime divisors. A prime ideal then would consist of all integers in $\varkappa$ which vanish at a given point on the Riemann surface. The separation of the points of the Riemann surface lying over a given point a in the X-plane is effected by the Puiseux expansions. The points at infinity, over $X = \infty$, are excluded. The elements of $\varkappa$ are the analytic functions on the closed Riemann surface which are regular everywhere except for isolated poles; the integers are those among them which have poles only in the points at infinity.

The question naturally arises whether one can algebraize the Puiseux expansions and thus develop a new universal method for the construction of prime divisors. In fact the oldest approach to the theory of algebraic numbers, that of Kummer, follows this line. Kummer did not win through to a perfectly general formulation, to whatever depths he penetrated in his special study of the cyclotomic fields. It was left to Hensel, who was guided by the analogy with the Puiseux expansions, to carry Kummer's approach to the finish by means of his idea of p-adic numbers. The next chapter will be devoted to a careful preparation and development of this method.

Chapter III

LOCAL PRIMADIC ANALYSIS (KUMMER-HENSEL)

1. Quadratic Number Field

The simplest field one can imagine over the field $\mathfrak{g}$ of common rational numbers is the "quadratic" field $\varkappa$ of degree 2. A quadratic equation is solved by extracting a square root. Hence we may choose as determining number of $\varkappa = \mathfrak{g}(\sqrt{a})$ the square root of a common integer $a \neq 1$ without squared factors, i.e., one containing no multiple prime factor. Let us first determine the integers among the numbers

$$\alpha = m + n\sqrt{a} \qquad (m, n \text{ rational}) \tag{1.1}$$

of $\varkappa$. The conjugate is

$$\alpha' = m - n\sqrt{a}.$$

The conditions for integrity are that the coefficients of the field equation, namely

$$\alpha + \alpha' = 2m, \qquad \alpha\alpha' = m^2 - an^2,$$

are rational integers. A fortiori

$$(2m)^2 - a(2n)^2 \tag{1.2}$$

and, on account of the first condition, $a(2n)^2$ must be integral. In view of a being "quadratfrei," this forces 2n to be integral, as one readily sees by prime factorization of the numbers concerned. If 2n is odd, we have

$$(2n)^2 \equiv 1 \pmod 4$$

and since (1.2) is $\equiv 0$ (4),

$$(2m)^2 \equiv a \quad (4).$$

The square of an integer is either $\equiv 0$ or $\equiv 1 \pmod 4$. The first case is here excluded because a is not divisible by the square 4. Hence we are left with the other alternative

$$a \equiv 1\ (4), \qquad 2m \text{ odd.}$$

The result is this: If $a \equiv 2$ or $\equiv 3 \pmod 4$ the integers of $\varkappa$ are of the form (1.1) with rational integers m, n; if $a \equiv 1\ (4)$ they are of the form

$$\frac{m + n\sqrt{a}}{2}$$

where the rational integers m and n are either both even or both odd,

$$m \equiv n \pmod 2)$$

In the first case

$$1 \text{ and } \theta = \sqrt{a}$$

form an integral basis, in the second case $1, \theta = \frac{1}{2}(-1 + \sqrt{a})$ form such a basis. By an integral basis of a field $\varkappa/k$ of degree n we understand a basis $\omega_1, \ldots, \omega_n$ consisting of integers such that every integer α in $\varkappa$ is expressible in the form

$$\alpha = a_1\omega_1 + \ldots + a_n\omega_n$$

where the components a_i of α are integers in k. The discriminant of the integral basis of $\mathfrak{g}(\sqrt{a})$ determined above

$$\begin{vmatrix} 1, & \theta \\ 1, & \theta' \end{vmatrix}^2 ,$$

is $d = 4a$ in the first, $d = a$ in the second case. Hence this discriminant is either $\equiv 0$ or $\equiv 1 \pmod 4$.

We next turn to the decomposition of a rational prime p into its prime divisors in $\varkappa$. Any prime divisor $\mathfrak{p}$ of $\varkappa$ goes into a rational prime number p. Writing $(p) = \mathfrak{p}\mathfrak{n}$ we obtain for the norm

$$p^2 = Nm\,\mathfrak{p} \cdot Nm\,\mathfrak{n}.$$

Hence there are two possibilities:

(1) $\mathrm{Nm}\,\mathfrak{p} = p^2$, $\mathrm{Nm}\,\mathfrak{r} = 1$; $\mathfrak{r} = (1)$, $\mathfrak{p} = (p)$;

(2) $\mathrm{Nm}\,\mathfrak{p} = p$ or $(p) = \mathfrak{p}\mathfrak{p}'$.

Either p itself is prime in κ or it splits into two conjugate prime divisors $\mathfrak{p}$, $\mathfrak{p}'$.

The second case can happen only, and will happen, if there exists an integer α not divisible by p such that $(p,\alpha) \neq (1)$. Then $\mathfrak{p} = (p,\alpha)$ is the one and $(p,\alpha') = \mathfrak{p}'$ the other prime factor. The following equation involving the indeterminate z,

$$\mathrm{Nm}(pz + \alpha) = p^2z^2 + (\alpha + \alpha')pz + \mathrm{Nm}\,\alpha$$

shows that

$$(p,\alpha) \neq (1) \quad \text{if} \quad \mathrm{Nm}\,\alpha : p.$$

Hence p will split if and only if there is a number α which itself is not, but whose norm is, divisible by p.

We first consider the case

$$a \equiv 2 \text{ or } 3 \pmod 4.$$

With the notation (1.1) the congruence

$$\mathrm{Nm}\,\alpha = m^2 - an^2 \equiv 0 \ (p)$$

must have an integral solution

$$(m,n) \not\equiv (0,0) \pmod p.$$

$n : p$ is thereby excluded because $n : p$ would imply $m^2 : p$ and hence $m : p$. We must therefore require

$$x^2 \equiv a \ (p)$$

to have a solution $x = b$, and then

$$\mathfrak{p} = (p,\ b + \sqrt{a}) \quad \text{and} \quad \mathfrak{p}' = (p,\ b - \sqrt{a})$$

are the two prime factors of p. For $p = 2$ one gets

$$(2) = \mathfrak{r}^2$$

with

$$\mathfrak{r} = (2, \sqrt{a}) \quad \text{if} \quad a \equiv 2 \ (4),$$

$$\mathfrak{r} = (2, 1 - \sqrt{a}) \quad \text{if} \quad a \equiv 3 \ (4).$$

If p is odd and a factor of a, then

$$(p) = \mathfrak{p}^2 \quad \text{with} \quad \mathfrak{p} = (p, \sqrt{a}).$$

If p is odd and not a factor of a, then p splits into two (distinct) factors or does not split at all according to whether or not a is quadratic residue mod p. Using Gauss' quadratic residue symbol, we have found

$$(p) = \mathfrak{p}\mathfrak{p}' \text{ in } \mathfrak{g}(\sqrt{a}) \quad \text{if} \quad (\frac{a}{p}) = +1,$$

$$(p) = \mathfrak{p} \qquad\qquad \text{if} \quad (\frac{a}{p}) = -1.$$

We now turn to the other case $a \equiv 1$ (4).

$$\alpha = m + n \frac{-1 + \sqrt{a}}{2},$$

$$\text{Nm } \alpha = (m - \frac{n}{2})^2 - \frac{an^2}{4} = m^2 - mn + \frac{1 - a}{4} \cdot n^2.$$

Again, decomposition takes place if $\text{Nm } \alpha \equiv 0$ (p) has a solution in integers m, n with $n \not\equiv 0$ (p), i.e., if the congruence

$$(x - \frac{1}{2})^2 - \frac{a}{4} = x^2 - x + \frac{1 - a}{4} \equiv 0 \ (p)$$

is solvable.

p = 2 splits (into two distinct factors) or does not according as $\frac{1 - a}{4}$ is even or odd,

$$a \equiv 1 \quad \text{or} \quad a \equiv 5 \ (\text{mod } 8).$$

p odd and factor of a:

$$p = \mathfrak{p}^2 \quad \text{with} \quad \mathfrak{p} = (p, \sqrt{a}).$$

p odd and not factor of a: p splits (into two distinct factors) if

$$(2x - 1)^2 \equiv a \ (4p)$$

is solvable or if $x^2 \equiv a$ (p) has an odd solution. [$x^2 \equiv a$ (4) is then automatically satisfied.] But if the last congruence modulo p has any solution b it also has an odd solution, namely either b or b + p.

We may summarize the results in both cases as follows:

Theorem III 1, A.

(i) If p is a divisor of the discriminant d, then $(p) = \mathfrak{p}^2$.

(ii) If p is odd and not a divisor of d, it splits into two distinct prime factors or does not split according as

$$\left(\frac{a}{p}\right) = +1 \quad \text{or} \quad -1.$$

(iii) If p = 2 is not a divisor of d (then $a \equiv 1$ *(4) and) the same alternative holds according to the two cases*

$$a \equiv 1 \quad \text{or} \quad \equiv 5 \quad (\text{mod } 8).$$

The determining equation of the field is

$$x^2 - a = 0. \tag{1.3}$$

One finds that the question, does or does not p split, is decided by the same alternative for the congruence mod p corresponding to the equation (1.3):

$$x^2 - a \equiv 0 \ (p).$$

Only p = 2 is an exception. This is Kummer's basic observation which he carried over to other fields. However, such exceptions as p = 2 in the quadratic case raised an obstacle which he was unable to overcome entirely. In the next two sections we are going to study a situation of some generality in which this difficulty does not arise.

2. Kummer's Theory: Decomposition

The ground field k is supposed to be a Dedekind field, and the field $\varkappa$ over k of degree n to possess a determining number θ which is an integer and such that every integer

$$\alpha = c_0 + c_1\theta + \dots + c_{n-1}\theta^{n-1}$$

has integral components c_i in k.

The defining equation is denoted by

$$F(x) = x^n + a_1x^{n-1} + \dots + a_n \quad (a_i \text{ integers}),$$

$F(\theta) = 0$. Let $\mathfrak{p}$ be a prime divisor in k.

Lemma III 2, A. $c_0 + c_1\theta + \dots + c_{n-1}\theta^{n-1} \equiv 0\ (\mathfrak{p})$ *only if all integers c_i in k are $\equiv 0(\mathfrak{p})$.*

Proof. The lemma is trivial if k is the rational ground field $\mathfrak{g}$. For then $\mathfrak{p}$ is a rational prime number p and the hypothesis that

$$\frac{c_0}{p} + \frac{c_1}{p}\theta + \dots + \frac{c_{n-1}}{p}\theta^{n-1}$$

be an integer in $\varkappa$ implies that its coefficients $\frac{c_i}{p}$ are integers in k. A slight modification of this obvious argument is needed for a k of more general nature.

In k we choose what we shall call a prime number to $\mathfrak{p}$, namely an integer p divisible by $\mathfrak{p}$ but not by $\mathfrak{p}^2$. $(p) = \mathfrak{p}\mathfrak{a}$. Let b be an integer divisible by $\mathfrak{a}$ but not by $\mathfrak{p}\mathfrak{a}$. Then $(b) = \mathfrak{a}\mathfrak{b}$, $\mathfrak{b}$ not divisible by $\mathfrak{p}$; hence b is not divisible by $\mathfrak{p}$.

$$\frac{c_0 b}{p} + \frac{c_1 b}{p}\theta + \dots + \frac{c_{n-1}b}{p}\theta^{n-1}$$

is integral in $\varkappa$; consequently the coefficients are integral in k and this shows c_i to be $: \mathfrak{p}$.

Let $\mathfrak{P}$ be a prime factor of $\mathfrak{p}$ in $\varkappa$ and

$$l_0 + l_1\theta + \dots + l_f\theta^f \equiv 0\ (\mathfrak{P})$$

the congruence of lowest degree f mod $\mathfrak{P}$ with integral coefficients l_i in k that are not all $\equiv 0\ (\mathfrak{p})$. In particular, $l_f \not\equiv 0\ (\mathfrak{p})$, and since the residues mod $\mathfrak{p}$ form a field $k_\mathfrak{p}$ we may assume $l_f = 1$. Our polynomials like

$$L(x) = x^f + l_{f-1}x^{f-1} + \dots + l_1x + l_0$$

are looked upon as polynomials in $k_\mathfrak{p}$. The residue field

$\kappa_{\mathfrak{P}}$ is of degree f over $k_{\mathfrak{p}}$. Every polynomial G(x) in $k_{\mathfrak{p}}$ with the property

$$G(\theta) \equiv 0 \ (\mathfrak{P}) \tag{2.1}$$

is divisible by L(x) in $k_{\mathfrak{p}}$:

$$G(x) \equiv L(x) \cdot G^*(x) \quad (\mathfrak{p}).$$

(Otherwise the remainder of the division G(x) : L(x) which takes place in the coefficient field $k_{\mathfrak{p}}$ would yield a congruence of the sort (2.1) for θ of lower degree than f.) In particular

$$F(x) \equiv L(x) \cdot H(x) \quad (\mathfrak{p}).$$

I maintain that

$$\mathfrak{P} = (\mathfrak{p}, L(\theta)). \tag{2.2}$$

The right side $\mathfrak{A}$ certainly is divisible by $\mathfrak{P}$. Each integer α is

$$\equiv c_0 + c_1\theta + \dots + c_{f-1}\theta^{f-1} \ (\mathfrak{A}).$$

This expression yields a complete residue system mod $\mathfrak{P}$ if each of the coefficients c_i ranges in k over a full residue system mod $\mathfrak{p}$; it is $\equiv 0 \ (\mathfrak{P})$ only if all coefficients c_i are $\equiv 0 \ (\mathfrak{p})$. Thus a number $\not\equiv 0 \ (\mathfrak{A})$ is $\not\equiv 0 \ (\mathfrak{P})$, or any integer divisible by $\mathfrak{P}$ is divisible by $\mathfrak{A}$; hence $\mathfrak{A} = \mathfrak{P}$.

Let us suppose $\mathfrak{p}$ to be exactly divisible by the power $\mathfrak{P}^e$. If $e > 1$ then $L(\theta)$ is divisible by the first power of $\mathfrak{P}$ only, therefore $H(\theta) \equiv 0 \ (\mathfrak{P})$ and H(x) again contains the factor L(x). Consequently F(x) will contain the factor $(L(x))^e$.

Let

$$\mathfrak{p} = \mathfrak{P}_1^{e_1} \dots \mathfrak{P}_g^{e_g}$$

be the decomposition of $\mathfrak{p}$ in κ into powers of distinct prime divisors. The polynomials $L_1(x), L_2(x), \dots$ in $k_{\mathfrak{p}}$ are determined for $\mathfrak{P}_1, \mathfrak{P}_2, \dots$ as L(x) was for $\mathfrak{P}$.

$$F(x) \equiv L_1^{e_1}(x) \cdot H_1(x) \quad (\mathfrak{p}).$$

On account of (2.2) we find

$$H_1(\theta) \equiv 0 \qquad (\mathfrak{p}_2^{e_2}\mathfrak{p}_3^{e_3} \dots),$$

and repeating our argument for $\mathfrak{p}_2$ etc., finally

$$(2.3) \qquad F(x) \equiv L_1^{e_1}(x) \dots L_g^{e_g}(x) \cdot F^*(x) \quad (\mathfrak{q}).$$

The congruence

$$L_1^{e_1}(\theta) \dots L_g^{e_g}(\theta) \equiv 0$$

holds modulo $\mathfrak{p}_1^{e_1} \dots \mathfrak{p}_g^{e_g} = \mathfrak{q}$, and thus the lemma forbids $L_1^{e_1} \dots L_g^{e_g}$ to be of a degree less than n in $k_{\mathfrak{q}}$. Therefore it is of degree n,

$$(2.4) \qquad e_1f_1 + \dots + e_gf_g = n,$$

$F^*(x) = 1$ in $k_{\mathfrak{q}}$ and (2.3) changes into

$$(2.5) \qquad F(x) \equiv L_1^{e_1}(x) \dots L_g^{e_g}(x) \quad (\mathfrak{q}).$$

$L_1, \dots, L_g$ are distinct polynomials in $k_{\mathfrak{q}}$. Were $L_1 = L_2$ then $L_1(\theta)$ would be divisible by $\mathfrak{p}_1\mathfrak{p}_2$, contrary to (2.2).

Hence this final and beautiful proposition which is due to Kummer:

Theorem III 2, B. *The decomposition of $\mathfrak{q}$ in $\varkappa$ runs parallel in every respect to the decomposition of $F(x)$ in $k_{\mathfrak{q}}$.*

The results of §1 are a simple corollary thereof.

One point is still to be cleared up. We have introduced a "degree" f of $\mathfrak{p}$ which is the degree of $\varkappa_{\mathfrak{p}}/k_{\mathfrak{q}}$. In view of the two equations (II, 8.3), (III, 2.4) it is to be guessed that this Kummer degree f coincides with the Kronecker degree f' defined by

$$\text{Nm}\,\mathfrak{p} = \mathfrak{q}^{f'}.$$

We multiply L(x) with the consecutive powers $1, x, \dots, x^{n-1}$ of the variable:

$$\left.\begin{aligned} L(x) \cdot 1 &\equiv a_{11} \cdot 1 + \dots + a_{1n}x^{n-1}, \\ &\dots\dots\dots\dots \\ L(x) \cdot x^{n-1} &\equiv a_{n1} \cdot 1 + \dots + a_{nn}x^{n-1} \end{aligned}\right\} \quad (\text{mod. } F(x)).$$

When we substitute θ for x we find that owing to the very definition of norm, the norm of $\pi = L(\theta)$ equals $|a_{ik}|$. We use $1, x, \ldots, x^{n-1}$ as indicators of the several rows and then can add to each of the last f rows suitable multiples of the previous ones such that their indicators $x^{n-f}, \ldots, x^{n-1}$ are changed into

$$H(x) \cdot 1, \ldots, H(x) \cdot x^{f-1}.$$

If we now make use of the fact

$$F(x) \equiv L(x) \cdot H(x) \quad (\mathfrak{p})$$

we see that all elements of the modified f last rows become divisible by $\mathfrak{p}$. Hence $\mathrm{Nm}\ \pi : \mathfrak{p}^f$. If $\mathfrak{p} = (p_1, \ldots, p_r)$ the same remains true for

$$\pi + p_1 z_1 + \ldots + p_r z_r \qquad (z \text{ indeterminates})$$

instead of π. Thus $\mathrm{Nm}\ \mathfrak{P} : \mathfrak{p}^f$, or

Kronecker degree $f' \geqq$ Kummer degree f.

As this is true for $\mathfrak{P}_1, \ldots, \mathfrak{P}_g$ and

$$e_1 f_1 + \ldots + e_g f_g = n, \qquad e_1 f_1' + \ldots + e_g f_g' = n,$$

the equality sign must prevail for all g divisors $\mathfrak{P}$.

3. Kummer's Theory: Discriminant

The case under consideration is characterized by the existence of an integral basis of the form

$$1, \theta, \ldots, \theta^{n-1}.$$

Its discriminant d shall be called the discriminant of the field.

$$d = \pm \mathrm{Nm}\ \delta, \qquad \delta = \dot{F}(\theta).$$

$$\dot{F}(x) = e_1 L_1^{e_1-1} \dot{L}_1 \cdot L_2^{e_2} L_3^{e_3} \ldots + e_2 L_2^{e_2-1} \dot{L}_2 \cdot L_1^{e_1} L_3^{e_3} \ldots + \ldots$$

proves that δ is divisible by $\mathfrak{P}_1^{e_1-1}$ and by no higher power, unless either $e_1 1$ or $\dot{L}_1(\theta) : \mathfrak{P}_1$.

Theorem III 3, A. *If $\mathfrak{p} : \mathfrak{P}^e$, then the differential* δ *is divisible by* $\mathfrak{P}^{e-1}$. *If* $\mathfrak{p} = \mathfrak{P}_1^{e_1}\mathfrak{P}_2^{e_2}\dots$ *then the discriminant d is divisible by*

$$\mathfrak{p}^{(e_1-1)f_1 + (e_2-1)f_2 + \dots}$$

The exponents are exact, unless $\varkappa_{\mathfrak{P}}/k_{\mathfrak{p}}$ *is not separable, or el* : $\mathfrak{p}$. *Both circumstances will definitely increase the exponent.*

(When we speak of el being divisible by $\mathfrak{p}$ we mean of course that the multiple el of the unit 1 in k is divisible by $\mathfrak{p}$.)

I propose to call k a numerical field provided $k_{\mathfrak{p}}$ is strictly finite for every prime divisor $\mathfrak{p}$. The number of residues then is a power $P = p^{f_0}$ of a certain rational prime number p. $\varkappa_{\mathfrak{P}}$ is also strictly finite, and consists of P^f residues. According to the theory of strictly finite fields, $\varkappa_{\mathfrak{P}}/k_{\mathfrak{p}}$ is a Galois field of degree f with a cyclic Galois group.

$$1 : \theta \to \theta, \qquad s : \theta \to \theta^P, \dots, \; s^{f-1} : \theta \to \theta^{P^{f-1}}$$

are its distinct automorphisms, while $s^f = 1$, i.e., $\alpha^{P^f} \equiv \alpha(\mathfrak{P})$ for every integer α in $\varkappa$. Hence:

Theorem III 3, B. *In the case of a numerical field k the degree f is the least exponent satisfying the congruence*

$$\theta^{P^f} \equiv \theta(\mathfrak{P}),$$

and

$$L(x) \equiv (x - \theta)(x - \theta^P)\dots(x - \theta^{P^{f-1}}) \quad (\mathfrak{P}).$$

4. Prime Cyclotomic Fields

Suppose l is a rational prime number; we study with Kummer the l-cyclotomic field $k = \mathfrak{R}(\zeta)$ over the ground field $\mathfrak{R}$ of rational numbers with the determining equation

$$F(x) = x^{l-1} + \dots + x + 1 \left\{ = (x - \zeta)(x - \zeta^2)\dots(x - \zeta^{l-1}) \right\}.$$

We have found that the principal divisor

$$\mathfrak{l} = (\lambda), \qquad \lambda = 1 - \zeta,$$

is prime and of degree l, and that l splits according to

$$(l) = \ell^{l-1}.$$

Let us form the differential of ζ:

$$\dot{F}(\zeta) = (\zeta - \zeta^2) \dots (\zeta - \zeta^{l-1}) = \zeta^{l-2}(1 - \zeta) \dots (1 - \zeta^{l-2}).$$

By means of

$$\mathrm{Nm}(1 - \zeta) = (1 - \zeta) \dots (1 - \zeta^{l-1}) = l, \qquad \mathrm{Nm}(-\zeta) = 1,$$

we find

$$\dot{F}(\zeta) = \frac{l \cdot \zeta^{l-2}}{1 - \zeta^{l-1}} = \frac{l}{-\zeta(1 - \zeta)},$$

$$\mathrm{Nm}\ \dot{F}(\zeta) = l^{l-1}/l = l^{l-2}.$$

Fitted with the correct sign, this norm is the discriminant of the basis $1, \zeta, \dots, \zeta^{l-2}$ which therefore has no other prime factor than l. We maintain that our basis is an integral basis, that is to say:

Theorem III 4, A.

$$c_0 + c_1\zeta + \dots + c_{l-2}\zeta^{l-2}$$

is integral in k if and only if the coefficients c_i are rational integers.

The proof depends on a general proposition about a field κ/k of degree n and a basis $\omega_1, \dots, \omega_n$ consisting of integers.

Lemma III 4, B. *The components a_i of an integer α in κ,*

$$(4.1) \qquad \alpha = a_1\omega_1 + \dots + a_n\omega_n \qquad (a_i \text{ in } k)$$

are of the form: integers/discriminant $d(\omega_1 \dots \omega_n)$.

Indeed (4.1) implies the equations

$$S(\alpha\omega_i) = \sum_k S(\omega_i\omega_k) \cdot a_k,$$

and as $S(\alpha\omega_i)$, $S(\omega_i\omega_k)$ are integral, the solution of this set of linear equations by means of determinants yields it in the form of a quotient whose numerator is integral and whose denominator is

$$|S(\omega_i\omega_k)| = d.$$

Thus we are sure that in our case any integer is of the form

$$\frac{c_0 + c_1\zeta + \ldots + c_{l-2}\zeta^{l-2}}{l^{l-2}}$$

or by using $\lambda = 1 - \zeta$ instead of ζ,

$$\frac{c_0 + c_1\lambda + \ldots + c_{l-2}\lambda^{l-2}}{l^{l-2}}$$

(c_i rational integers).

One after the other of the factors l of the denominator may be divided into the coefficients c of the numerator, because of the following lemma to which our theorem is thus reduced:

Lemma III 4, C. *If*

(4.2) $$c_0 + c_1\lambda + \ldots + c_{l-2}\lambda^{l-2}$$

with rational integral coefficients c_i is divisible by l, then each of the coefficients is divisible by l.

Proof. $l \sim \lambda^{l-1}$. The hypothesis at once leads to $c_0 : \mathcal{l}$. Hence c_0 and l cannot be relative prime and therefore $c_0 : l$. Subtracting c_0 from (4.2) one finds

$$c_1 + c_2\lambda + \ldots + c_{l-2}\lambda^{l-3} \quad \text{to be divisible by } \mathcal{l}^{l-2},$$

therefore $c_1 : \mathcal{l}$, $c_1 : l$; and so on.

This once accomplished, the road is free for the application of the theory of the last two sections. Let p be a rational prime number $\neq l$ and $\mathfrak{p}$ a prime divisor of p in k. If

$$p^f \equiv 1 \quad (l)$$

then

$$\zeta^{p^f} = \zeta, \quad \text{a fortiori} \quad \zeta^{p^f} \equiv \zeta \pmod{\mathfrak{p}}.$$

However, if $p^f \not\equiv 1 \; (l)$, then

$$\zeta^{p^f} - \zeta = \zeta(\zeta^{p^f - 1} - 1) \sim \lambda$$

is prime to $\mathfrak{p}$. Thus we obtain:

Theorem III 4, D. *l decomposes according to the equation $(l) = \mathfrak{l}^{l-1}$ into $l - 1$ equal prime divisors of degree 1.*

A rational prime $p \neq l$ splits into a number g of distinct prime divisors $\mathfrak{p}_1, \ldots, \mathfrak{p}_g$ of the same degree f,

$$(p) = \mathfrak{p}_1 \cdots \mathfrak{p}_g.$$

f is the least exponent for which

$$p^f \equiv 1 \pmod{l}$$

and

$$f \cdot g = l - 1.$$

(Because of Fermat's formula

$$p^{l-1} \equiv 1 \pmod{l}$$

f must indeed be a divisor of $l - 1$.)

The decomposition of $F(x)$ runs parallel: $F(x)$ splits into incongruent mod p irreducible factors $L(x)$ in $k_{\mathfrak{p}}$ of degree f,

$$x^{l-1} + \ldots + x + 1 \equiv L_1(x) \ldots L_g(x) \pmod{p}.$$

$$\mathfrak{p}_j = (p, L_j(\zeta)).$$

$$L_j(x) \equiv (x - \zeta)(x - \zeta^p) \ldots (x - \zeta^{p^{f-1}}) \pmod{\mathfrak{p}_j}.$$

5. Program

We now follow up the analogy of Kummer's theory with the analytic theory of rational and algebraic functions of a single variable X. Gradually analogy will give way to identity.

(I) In the field $k = \Omega(X)$ the linear expression $X - a$ is a prime. Any integer (polynomial) $F = F(X)$ is congruent to a constant $F(a)$ modulo this prime. Hence the constants yield a complete system of residues. A rational function without a pole at $X = a$ may be written as a fraction of two polynomials F/G where $G(a) \neq 0$, and developed by powers of $X - a$:

$$c_0 + c_1(X - a) + c_2(X - a)^2 + \dots .$$

We do not look upon this expansion as a convergent series which (in a certain domain of the X-plane) gives the numerical values of the rational function, but rather as an analysis of its behavior modulo higher and higher powers of $X - a$. In other words we interpret our equation as indicating the sequence of congruences

$$F(X) \equiv \left\{c_0 + c_1(X - a) + \dots + c_{\nu-1}(X - a)^{\nu-1}\right\} \cdot G(X) \pmod{(X - a)^\nu}$$

$$[\nu = 1, 2, \dots].$$

In fact the recurrent computation of the coefficients c_ν takes place in this fashion. Take as an example the expansion

$$\frac{1}{1 - X} = 1 + X + X^2 + \dots$$

which for us means

$$1 \equiv (1 - X)(1 + X + \dots + X^{\nu-1}) \pmod{X^\nu}.$$

This simple change of attitude algebraizes the function-theoretic expansions and is the decisive step of rapprochement between the two seemingly hostile camps.

What is the analogous procedure in the field $\mathcal{G}$ of rational numbers? With respect to a prime p we use

$$(5.1) \qquad 0, 1, \dots, p - 1$$

as a full system of residues. For any rational fraction $\frac{a}{b}$ whose denominator b is not divisible by p we can construct a unique expansion

$$\frac{a}{b} = c_0 + c_1 p + c_2 p^2 + \dots$$

with coefficients c taken from the residue set (5.1) in the sense that

$$(5.2) \qquad a \equiv b(c_0 + c_1 p + \dots + c_{\nu-1} p^{\nu-1}) \pmod{p^\nu}$$

$$[\nu = 1, 2, \dots].$$

This is what Hensel calls the p-adic expansion of the fraction. There is no question of convergence, but the series describes the arithmetic behavior of the number a/b modulo p and all its powers. Indeed, if $c_0, c_1, \dots, c_{\nu-1}$ are so constructed as to satisfy (5.2) the next coefficient c_ν is uniquely determined by the solvable congruence

$$bc_\nu \equiv \frac{a - b(c_0 + \dots + c_{\nu-1} p^{\nu-1})}{p^\nu} \pmod{p}.$$

In using the notation customary for decimal fractions we say with Hensel that a/b has the p-adic expansion, or is equal to,

$$. c_0 c_1 c_2 \dots$$

in the realm of p ("im Bereich von p"), and write

$$\frac{a}{b} = . c_0 c_1 c_2 \dots \quad (p).$$

For instance

$$\frac{1}{12} = .3626262 \dots \quad (7).$$

Incidentally the p-adic expansions of rational numbers are periodic, for the same reason as their decimal expansions.

The limitation imposed by $b \not\equiv 0$ (p) can be removed if we allow expansions of the form

$$c_{-h} p^{-h} + \dots + c_{-1} p^{-1} + c_0 + c_1 p + \dots$$

including a finite number of terms with negative exponents--just as in the theory of rational functions we must admit expansions of the type

$$\frac{c_{-h}}{(X - a)^h} + \dots + \frac{c_{-1}}{X - a} + c_0 + c_1(X - a) + \dots .$$

Examples:

$$\frac{1}{X - X^2} = \frac{1}{X} + 1 + X + X^2 + \dots .$$

$$\frac{1}{84} = 3 \cdot 7^{-1} + 6 + 2 \cdot 7 + 6 \cdot 7^2 + 2 \cdot 7^3 + \dots \qquad (7)$$

(II) When we turn from rational to algebraic functions of X we start with an equation

$$(5.3) \qquad \theta^n + f_1(X)\theta^{n-1} + \dots + f_n(X) = 0$$

whose coefficients f_i are polynomials. Let us consider a point X = a in the X-plane over which the Riemann surface $\mathcal{F}$ has n separate sheets. In the neighborhood of that point we have n roots of (5.3) described by power series of X - a:

$$\theta^{(1)} = b_0^{(1)} + b_1^{(1)}(X - a) + b_2^{(1)}(X - a)^2 + \dots ,$$
$$\dots\dots\dots\dots\dots\dots\dots\dots\dots$$
$$\theta^{(n)} = b_0^{(n)} + b_1^{(n)}(X - a) + b_2^{(n)}(X - a)^2 + \dots$$

which represent the function θ on $\mathcal{F}$ in the neighborhoods of the n points $\mathcal{Y}_1, \dots, \mathcal{Y}_n$ lying over a. The functional elements $\theta^{(1)}$, $\theta^{(2)}$ are certainly different, the points of the Riemann surface are so defined as to correspond to the elements of the analytic function in Weierstrass' theory; but it may well come to pass that $b_0^{(1)} = b_0^{(2)} = b_0$, i.e., that the values of θ coincide at two distinct points $\mathcal{Y}_1$, $\mathcal{Y}_2$ (at which X also takes on the same value a). To be sure, any function of our algebraic field is expressible by X and θ in a rational fashion. However, this has not the consequence that every function α assumes equal values of $\mathcal{Y}_1$ and $\mathcal{Y}_2$ although

$$X(\mathcal{Y}_1) = X(\mathcal{Y}_2) \quad \text{and} \quad \theta(\mathcal{Y}_1) = \theta(\mathcal{Y}_2).$$

For instance if $b_1^{(1)} \neq b_1^{(2)}$ then

$$\frac{\theta - b_0}{X - a}$$

takes on the two different values $b_1^{(1)}$ and $b_2^{(1)}$ at $\mathcal{Y}_1$ and $\mathcal{Y}_2$. In order to separate the points of the Riemann surface, one must consider either the values of all functions of the field or, if one operates with θ alone, the whole expansion of θ. We decide in favor of the second procedure, and this means that we study θ not only mod. $X - a$, but "in the realm of the rational prime $X - a$."

If we hit upon a point $X = a$ over which lies a ramification point $\mathfrak{y}$ with e sheets, one uses the local parameter

$$\tau = (X - A)^{1/e}$$

and in the neighborhood of $\mathfrak{y}$ will obtain an expansion

$$\theta = b_0 + b_1\tau + b_2\tau^2 + \dots .$$

It yields e roots $\theta^{(1)}$ ..., $\theta^{(e)}$ since τ may be replaced by $\zeta\tau$ where ζ is any of the eth roots of unity. The polynomial

$$(Y - \theta^{(1)}) \dots (Y - \theta^{(e)})$$

in Y has coefficients which are regular power series in $X - a$. The points $\mathfrak{y}_1, \dots, \mathfrak{y}_g$ over a with the ramification orders $e_1, \dots, e_g$ will therefore correspond to the decomposition of (5.3) into irreducible factors, if we operate not in the field of rational functions of X, but in the field of power series of $X - a$. The decomposition is carried through only locally, in the neighborhood of $X = a$. One has

$$(5.4) \qquad e_1 + \dots + e_g = n.$$

The rational prime function $X - a$ vanishes at $\mathfrak{y}_1, \dots, \mathfrak{y}_g$ with the respective orders $e_1, \dots, e_g$ and nowhere else, a fact which we express by

$$(5.5) \qquad X - a \sim \mathfrak{y}_1^{e_1} \dots \mathfrak{y}_g^{e_g},$$

There exists a function π in our function field which vanishes at $\mathfrak{y} = \mathfrak{y}_1$ with exactly the first order. It is in better keeping with our algebraic tendencies if we employ such a π rather than τ as the local parameter.

Is it now clear wherein Kummer's theory erred? The defining equation

$$f(\theta) = a_0\theta^n + a_1\theta^{n-1} + \dots + a_n = 0$$

for $\varkappa = \mathfrak{g}(\theta)$ over the ground field $\mathfrak{g}$ should not be decomposed merely mod p, but in the realm of the rational prime number: this decomposition will correspond to the separation of the prime divisors of $\varkappa$ which go into p. To carry out this idea we operate in the field of all p-adic numbers, i.e., of all formal series

$$c_h p^h + c_{h+1} p^{h+1} + \dots \tag{5.6}$$

[h any integer, c_h, c_{h+1}, ... any residues taken from the set (5.1)], just as in the case of algebraic functions we operate in the field of all formal power series

$$c_h(X - a)^h + c_{h+1}(X - a)^{h+1} + \dots$$

of $X - a(c_h, c_{h+1}, \dots$ any constants).

Comparison of (5.4) with (2.4) indicates that in the theory of algebraic functions of the variable X all prime divisors are of degree $f = 1$. The reason for this is that our coefficient field is the field Ω of all complex numbers. This peculiarity would disappear if Ω be replaced by a coefficient field that is not algebraically closed.

(III) Before going on, let us practice the technique of factorizing polynomials

$$f(x) = a_0 x^n + \dots + a_n \tag{5.7}$$

with integral coefficients "in the realm of a prime p," and once more start with the quadratic polynomial

$$f(x) = x^2 - a.$$

If $p \neq 2$ and a prime to p we decide by checking the numbers $x = 1, 2, \dots, p - 1$ whether or not a is quadratic residue, and in the first alternative we choose one of the two roots $\pm b$ (mod p),

$$x^2 - a \equiv (x - b)(x + b) \pmod{p}. \tag{5.8}$$

I maintain that the congruence

$$x^2 - a \equiv 0$$

if solvable mod p is also solvable modulo all higher powers of p: the factorization (5.8) can be uniquely pushed forward to the moduli $p^2, p^3, \dots$. Indeed, assuming that we we have ascertained a number b_ν such that

$$b_\nu^2 \equiv a \quad (\mathrm{mod}\ p^\nu)$$

we try to construct

$$b_{\nu+1} = b_\nu + p^\nu \cdot x$$

such that

$$b_{\nu+1}^2 \equiv a \quad (\mathrm{mod}\ p^{\nu+1}).$$

If $\nu \geqq 1$,

$$b_{\nu+1}^2 \equiv b_\nu^2 + 2b_\nu x \cdot p^\nu \quad (\mathrm{mod}\ p^{\nu+1}),$$

and hence we have merely to solve the linear congruence

$$2b_\nu x \equiv \frac{a - b_\nu^2}{p^\nu} \quad (\mathrm{mod}\ p)$$

which, because of $2b_\nu \not\equiv 0$ (p), has a unique solution x mod p. In this way the root $b_1 = b$ of the congruence mod p leads to a uniquely determined root

$$b + b'p + b''p^2 + \ldots$$

of the equation

$$x^2 - a = 0$$

in the realm of p. For example

$$\sqrt{11} = .224\ \ldots \quad (7).$$

The situation is slightly different for p = 2:

$$x^2 - a \equiv 0 \quad (2^\nu)$$

is solvable for any power ν of 2 if it is solvable for 2^3. Not from the first, but from the third power on, does the process here run along a smooth channel. Proof: a is odd. Suppose we are in possession of a b_ν such that

$$b_\nu^2 \equiv a \quad (2^\nu).$$

We try to find a

$$b_{\nu+1} = b_\nu + 2^{\nu-1} \cdot x$$

satisfying the congruence

$$b_{\nu+1}^2 \equiv a \quad (2^{\nu+1}).$$

If $\nu \geqq 3$, then

$$b_{\nu+1}^2 \equiv b_\nu^2 + 2^\nu b_\nu x \quad (2^{\nu+1}).$$

Hence

$$b_\nu x \equiv \frac{a - b_\nu^2}{2^\nu} \quad (2).$$

In other words we have to take $x \equiv 0$ or $1 \pmod 2$ according to whether the right member is even or odd.

This helps us to understand why 2 splits or does not split in the quadratic field $\mathcal{G}(\sqrt{a})$ with $a \equiv 1 \ (4)$ according to whether

$$a \equiv 1 \quad \text{or} \quad 5 \pmod 8):$$

1 is quadratic residue, 5 quadratic non-residue mod 8.

Our results concerning $x^2 - a$ are typical for any polynomial $f(x)$. If a factorization

$$(5.9) \qquad f(x) \equiv g(x) \cdot \bar{g}(x)$$

has been effected modulo a sufficiently high power p^λ of p it can be continued unambiguously to all subsequent powers $p^{\lambda+1}$, $p^{\lambda+2}$, ... In order to prove this and to fix the exponent λ explicitly, one must resort to the resultant $R(f,g)$ of two polynomials f, g, on which the elementary textbooks of algebra may be consulted. The discriminant $D(f)$ of f is the resultant $R(f,\dot{f})$. If

$$f = g \cdot \bar{g}$$

then

$$D(f) = D(g) \cdot D(\bar{g}) \cdot R^2(g,\bar{g}).$$

The congruence (5.9) modulo p^ν implies the corresponding congruence

$$(5.10) \qquad D(f) \equiv D(g) \cdot D(\bar{g}) \cdot R^2(g,\bar{g}) \qquad (\text{mod } p^\nu).$$

We assume f to be without multiple roots ($\varkappa/\mathfrak{g}$ to be separable), so that $D(f) \neq 0$. If the discriminant is exactly divisible by $p^{\lambda-1}$, then a factorization (5.10) with $\nu \geqq \lambda$ implies

$$R^2(g,\overline{g}) \not\equiv 0 \quad (p^\lambda),$$

or $R(g,\overline{g})$ is not divisible by a higher power than p^δ where

$$\delta = \frac{1}{2}\lambda - 1 \quad \text{if} \quad \lambda \text{ even},$$

$$\delta = \frac{1}{2}(\lambda - 1) \quad \text{if} \quad \lambda \text{ odd}.$$

In order to pass from the factorization mod p^ν, $\nu \geqq \lambda$,

$$f(x) \equiv g_\nu(x) \cdot \overline{g}_\nu(x) \qquad (\text{mod } p^\nu)$$

to the module $p^{\nu+1}$, we put

$$g_{\nu+1}(x) = g_\nu(x) + r_\nu(x) \cdot p^{\nu-\delta},$$

$$\overline{g}_{\nu+1}(x) = \overline{g}_\nu(x) + \overline{r}_\nu(x) \cdot p^{\nu-\delta},$$

where $r_\nu(x)$, $\overline{r}_\nu(x)$ are of the same formal degree as $g(x)$, $\overline{g}(x)$ respectively. We find

$$g_{\nu+1} \ \overline{g}_{\nu+1} \equiv g_\nu\overline{g}_\nu + (g_\nu\overline{r}_\nu + \overline{g}_\nu r_\nu)p^{\nu-\delta} \quad (p^{\nu+1})$$

and hence the desired result

$$g_{\nu+1}(x) \cdot \overline{g}_{\nu+1}(x) \equiv f(x) \quad (p^{\nu+1})$$

if

$$g_\nu\overline{r}_\nu + \overline{g}_\nu r_\nu = \frac{f - g_\nu\overline{g}_\nu}{p^\nu} \cdot p^\delta.$$

These are linear equations for the unknown coefficients of r_ν and $\overline{r}_\nu$. Since their determinant $R(g_\nu,\overline{g}_\nu)$ is not divisible by a higher power than p^δ and the coefficients of the right members are all divisible by p^δ we obtain a solution r_ν, $\overline{r}_\nu$ with integral coefficients. In this way a factorization (5.9) for $\nu = \lambda$ can be uniquely continued to higher powers.

The discriminant of $x^2 - a$ is $4a$; therefore the exponent $\lambda = 1$ for $p \neq 2$ and $\lambda = 3$ for $p = 2$.

In Chapter I, §4, we studied this process. Given a polynomial $f(x)$, (5.7), of degree n in a given ground field k, we agreed to identify any two polynomials in k of an indeterminate θ if they are congruent mod. $f(\theta)$ and thus generated a commutative ring $\varkappa$ of degree n over k. If $f(x)$ is irreducible, then this ring is a field, but what happens in the general case? We assume $f(x)$ to be prime to its derivative $\dot{f}(x)$, so that it splits into distinct irreducible factors

$$f(x) = f_1(x) \dots f_g(x).$$

Each of them determines a simple extension $\varkappa_1, \dots, \varkappa_g$ of k, the degrees of which, $n_1, \dots, n_g$, are given by the degrees of the factors $f_1(x), \dots, f_g(x)$. I maintain that the ring $\varkappa$ is the direct sum of these fields; i.e., it is isomorphic to the ring of all g-tuples

$$(\alpha_1, \dots, \alpha_g)$$

whose members α_j vary independently, each over its field $\varkappa_j$, the addition and multiplication being defined by the rules

$$(\alpha_1, \dots, \alpha_g) + (\beta_1, \dots, \beta_g) = (\alpha_1 + \beta_1, \dots, \alpha_g + \beta_g),$$

$$(\alpha_1, \dots, \alpha_g) \cdot (\beta_1, \dots, \beta_g) = (\alpha_1\beta_1, \dots, \alpha_g\beta_g).$$

In fact, two polynomials $q(\theta)$ in k which are congruent mod. $f(\theta)$ are also congruent mod. $f_1(\theta), \dots,$ mod. $f_g(\theta)$. Hence each element α of $\varkappa$ is at the same time an element α_1 of $\varkappa_1, \dots, \alpha_g$ of $\varkappa_g$:

$$(5.11) \qquad \alpha \longrightarrow (\alpha_1, \dots, \alpha_g).$$

Vice versa, if $\alpha_1, \dots, \alpha_g$ are given elements in their respective fields $\varkappa_1, \dots, \varkappa_g$ then there is a single element α in $\varkappa$ such that (5.11) holds. Indeed, $q_1(x), \dots, q_g(x)$ being given polynomials, there exists a polynomial $q(x)$ such that

$$q(x) \equiv q_1(x) \pmod{f_1(x)}, \dots, q(x) \equiv q_g(x) \pmod{f_g(x)},$$

and $q(x)$ is uniquely determined modulo $f_1(x) \dots f_g(x) = f(x)$.

We apply this to our polynomial $f(x)$ in the field

$\mathfrak{P}(p)$ of p-adic numbers where f(x), though irreducible in the field $\mathfrak{P}$, breaks up into distinct irreducible factors

$$f(x) = f_1(x) \ldots f_g(x) \quad (p).$$

We have described before how this decomposition may be effected by a strictly finitistic way of calculation. The ring $\kappa(p)$ consists of all polynomials of θ mod. $f(\theta)$ whose coefficients are p-adic numbers. In writing them in the form (5.6) and ignoring for the moment the restrictions imposed upon the coefficients $c_h, c_{h+1}, \ldots,$ one sees that the elements of $\kappa(p)$ are formal expansions

$$\gamma_h p^h + \gamma_{h+1} p^{h+1} + \ldots$$

with coefficients γ in κ. Thus the description of $\kappa(p)$ becomes independent of the choice of the determining number θ.

When one cuts out a small circular neighborhood of the point X = a by a stamp cutting through the sheets of the Riemann surface which is spread over the X-plane, one gets something that falls apart into g pieces [of $e_1, \ldots, e_g$ sheets respectively; cf. the notation (5.5)], while the whole Riemann surface, owing to the irreducibility of the defining equation, consists of one piece. The corresponding algebraic event is the falling apart of the ring $\kappa(p)$ into the direct sum of g fields. On account of this analogy one refers to the investigation of the numbers of κ in the realm of p or its various prime divisors $\mathfrak{p}_1, \ldots, \mathfrak{p}_g$ as a local investigation at the prime spot p (or $\mathfrak{p}_1, \ldots, \mathfrak{p}_g$).

The preceding exposition shows a way of erecting the Kummer-Hensel theory independently of the Kronecker-Dedekind theory of divisors. Its basic notion is not that of a divisor, but of prime divisor or prime spot. However, here we want to prove that the falling apart of $\kappa(p)$ into fields corresponds to the decomposition of p into g powers of Kronecker prime divisors in κ. Hence we shall base our development of the p-adic method upon the ideas and results of Chapter II.

(IV) We have observed that in our algebraic theory the points of the Riemann surface at infinity (over X = ∞) have been omitted. But the theory of functions on a Riemann surface becomes much more harmonious if we include them, whereby the surface becomes a closed manifold. One is bound to ask oneself whether a similar procedure, the introduction of prime spots at infinity, is possible for other fields with a similar harmonizing effect. In general, this does not seem to be the case. However, there exists a deep-reaching analogy for the algebraic number fields proper, that is to say, the finite fields over $\mathfrak{g}$. Yet we do not propose to take up this suggestion before the next chapter.

With §6 we resume the systematic theory.

6. p-adic and $\mathfrak{p}$-adic Numbers

A rational number which may be written as a fraction

$$a/b \quad (a,\ b \text{ integers})$$

whose denominator is indivisible by the prime number p, is said to be locally integral at (the prime spot) p. The numbers which are integral at p form a ring. r is a local unit if both r and 1/r are local integers. A number which is locally integral everywhere (at every prime spot p) is absolutely integral.

If a system Σ of residues mod p is given, e.g.,

$$(6.1) \qquad \Sigma = (0,\ 1, \ldots,\ p - 1)$$

then there exists for any number c which is integral at p a residue c' in Σ such that $\frac{c - c'}{p}$ is integral at p; we write

$$c \equiv c' \quad (p).$$

Indeed, if

$$c = a/b; \quad a,\ b \text{ integers}, \quad b \text{ prime to } p,$$

one finds c' by solving the congruence

$$bc' \equiv a \quad (\text{mod } p).$$

In defining the p-adic numbers it is not convenient to restrict the coefficients c in (5.6) to the residues (6.1). First, the choice of this particular system of residues is somewhat arbitrary, at least far more arbitrary than the choice of constants as residues in the field $\Omega(X)$; and in an algebraic field over $\mathfrak{g}$ there is absolutely nothing to direct our choice. Secondly, the definition of addition and multiplication becomes much simpler without the restriction which would require the inclusion of that technique of "carrying over" and "borrowing" so familiar to us from the decimal system. Of course one has to pay for the abandonment of the residue system by an explicit definition of equality. Once one has gone so far, one may as well extinguish all traces of non-local concepts and admit as coefficients any rational numbers that are locally integral at p. Therefore the following set-up:

Definition (first part). *A sequence of numbers*

$$a_\nu \qquad (-\infty < \nu < \infty)$$

which are integral at p defines a p-adic number

$$A = \ldots a_{-2}a_{-1} \,.\, a_0a_1a_2 \ldots$$

provided all a_ν whose ν are sufficiently far to the left are zero, i.e., if there exists an h such that $a_\nu = 0$ for $\nu < h$.

We introduce the partial sums

$$A_\nu = \sum^{i=\nu-1} a_i p^i$$

which satisfy the congruences

$$A_{\nu+1} \equiv A_\nu \pmod{p^\nu}$$

whereby we mean that

$$\frac{A_{\nu+1} - A_\nu}{p^\nu} \; (= a_\nu)$$

is locally integral. A_ν describes the behavior of A modulo p^ν; in this sense we need not hesitate to write

$$A = \lim_{\nu \to \infty} A_\nu$$

(p-adic limit).

Definition (second part). *Two p-adic numbers*

$$A = \lim_{\nu \to \infty} A_\nu, \qquad B = \lim_{\nu \to \infty} B_\nu$$

are said to be equal if

$$A_\nu \equiv B_\nu \pmod{p^\nu}$$

for all ν.

This relationship, as it should be, is obviously reflexive, symmetric, and transitive. After having chosen a definite system Σ of residues mod p, there is among the p-adic numbers which equal a given one A, exactly one whose coefficients a_ν are in Σ; we call this the reduced form of A.

Addition. If the sequences a_ν, b_ν define the p-adic numbers A, B, the p-adic number defined by the sequence $a_\nu + b_\nu$ is called the sum A + B.

One readily sees that

$$A = A', \quad B = B' \quad \text{imply} \quad A + B = A' + B'.$$

Moreover, it is clear how substraction is carried out.

Multiplication. The product C = AB is defined by

$$c_\nu = \Sigma a_i b_k \quad (i + k = \nu).$$

Hence

$$C_\nu = \Sigma a_i p^i \cdot b_k p^k \quad (i + k < \nu)$$

(6.2) $$= \sum_{k<\nu} A_{\nu-k} \cdot b_k p^k.$$

It is essential that

(6.3) $$A = A', \quad B = B' \quad \text{imply} \quad AB = A'B'.$$

For C = AB and C' = A'B one derives from (6.2):

$$\frac{C_\nu - C'_\nu}{p^\nu} = \sum_{k<\nu} \frac{A_{\nu-k} - A'_{\nu-k}}{p^{\nu-k}} \cdot b_k,$$

hence

$$A = A' \quad \text{implies} \quad AB = A'B.$$

By double application of this result one arrives at (6.3).

$$A = a_0 + a_1 p + \dots$$

is a p-adic unit if a_0 is a local unit at p. We construct the inverse

$$1/A = B = b_0 + b_1 p + \dots$$

of the unit A by computing the coefficients $b_0, b_1, \dots$ recursively from the equations

$$a_0 b_0 = 1,$$

$$a_0 b_\nu + a_1 b_{\nu-1} + \dots + a_\nu b_0 = 0 \quad [\nu = 1, 2, \dots]$$

which at every step yield a local integer b_ν.

If the first coefficient a_h of the p-adic number

(6.4) $$A = a_h p^h + a_{h+1} p^{h+1} + \dots$$

is locally divisible by p, one can write A so as to have the coefficient of p^h vanish. Hence every p-adic number $A \neq 0$ may be written in such a way, (6.4), that the first coefficient a_h is not locally divisible by p. Then

$$A = p^h \cdot A' \quad \text{where} \quad A' = a_h + a_{h+1} p + \dots$$

is a p-adic unit. h is called the order of A, and A is said to be integral if $h \geqq 0$. We write

$$A \sim p^h.$$

The order satisfies the relations

$$\text{ord}\,(AB) = \text{ord}\,A \cdot \text{ord}\,B;$$

$$\text{ord}\,(A + B) \geqq \min\,(\text{ord}\,A, \text{ord}\,B).$$

The inverse B of A arises from the inverse

$$B' = b_0 + b_1 p + \dots$$

of A' by division with p^h:

$$B = b_0 p^{-h} + b_1 p^{-h+1} + \dots .$$

A being any number $\neq 0$, either A or 1/A is an integer; they are both integral if and only if A is a p-adic unit.

The p-adic numbers thus form a field $\mathfrak{g}(p)$, and it is clear in what sense the common rational numbers are contained in this field. More pedantically one may speak of an isomorphic mapping I_p of the numbers of $\mathfrak{g}$ into the p-adic field $\mathfrak{g}(p)$.

We have been careful to formulate our basic definitions so as to carry over immediately to any Dedekind field k. Since from now on we shall be concerned with Dedekind fields only, we adopt the term "ideal," which is customary in the literature, instead of divisor. Let $\mathfrak{p}$ be a prime ideal in k. A system Σ of residues is a system of integers such that every integer is congruent to one and only one of the integers in Σ. As basic number we use a prime number p to $\mathfrak{p}$, i.e., an integer divisible by $\mathfrak{p}$ but not by $\mathfrak{p}^2$. A number c is said to be integral at $\mathfrak{p}$ if it can be written as a fraction

$$a/b \qquad (a,\ b \text{ integers})$$

with a denominator b not divisible by $\mathfrak{p}$. There is then a uniquely determined residue c' in Σ such that $\frac{c - c'}{p}$ is integral at $\mathfrak{p}$. (Here the Dedekind nature of the field k comes in.) The local integers form a ring. The term local unit is employed as before.

Lemma III 6, A. *If the integer $b \neq 0$ is exactly divisible by $\mathfrak{p}^h$, then a/b is a local integer if and only if the integer a is divisible by $\mathfrak{p}^h$.*

Proof. If a/b is locally integral at $\mathfrak{p}$, then

$$a/b = a'/b' \quad \text{or} \quad ab' = a'b$$

with a denominator b' prime to $\mathfrak{p}$. By the last equation ab' and hence a is divisible by $\mathfrak{p}^h$.

Vice versa, suppose a to be divisible by $\mathfrak{p}^h$. We put $(p) = \mathfrak{p}\mathfrak{s}$ and determine an integer $s : \mathfrak{s}$ such that $(s)/\mathfrak{s} = \mathfrak{s}'$ is prime to $\mathfrak{p}$. Then $(s) = \mathfrak{s}\mathfrak{s}'$ is not divisible by $\mathfrak{p}$. Put

$$b' = \left(\frac{s}{p}\right)^h \cdot b, \qquad a' = \left(\frac{s}{p}\right)^h \cdot a.$$

Both are integers, b' is prime to $\mathfrak{p}$ and

$$a/b = a'/b'.$$

Corollary. *The local integer*

$$c = \frac{a}{b} \quad (a,\ b \text{ integers},\quad b \text{ prime to } \mathfrak{p})$$

is a local unit if and only if the numerator a is also prime to $\mathfrak{p}$.

Corollary. *c being a number* $\neq 0$, *either c or 1/c is locally integral at* $\mathfrak{p}$.

Proof. If c is written as a fraction a/b, and numerator and denominator are exactly divisible by $\mathfrak{p}^i$ and $\mathfrak{p}^h$ respectively, then the first alternative occurs if $i \geqq h$, the second if $h \geqq i$.

Corollary. *A number which is locally integral everywhere is absolutely integral.*

Proof. c = a/b, a and b integers. If

$$(b) = \mathfrak{p}_1^{e_1}\mathfrak{p}_2^{e_2} \ldots$$

is the factorization of the denominator into powers of distinct prime ideals, then the supposed local integrity of c at $\mathfrak{p}_1, \mathfrak{p}_2, \ldots$ requires a to be divisible by $\mathfrak{p}_1^{e_1}, \mathfrak{p}_2^{e_2}, \ldots$, hence by the product $\mathfrak{p}_1^{e_1}\mathfrak{p}_2^{e_2}, \ldots$, i.e., by b.

The $\mathfrak{p}$-adic numbers in k are introduced by means of the prime number p to $\mathfrak{p}$ as formal sums

$$\sum_{\nu=-\infty}^{+\infty} a_\nu p^\nu$$

whose coefficients a_ν break off toward the left side $\nu \to -\infty$. All the above definitions carry over, and hence these $\mathfrak{p}$-adic numbers form a field $k(\mathfrak{p})$ containing k itself. The only new remark is to the effect that this field is independent of the choice of the basic number p. Indeed, if p^* is another prime number to $\mathfrak{p}$, then $\frac{p}{p^*}$ is a local unit at $\mathfrak{p}$, and the convention

$$\Sigma a_\nu p^\nu = \Sigma a_\nu^*(p^*)^\nu \quad \text{with} \quad a_\nu^* = a_\nu\left(\frac{p}{p^*}\right)^\nu$$

identifies with one another the $\mathfrak{p}$-adic numbers of the bases p and p^*. This correspondence obviously leaves undisturbed equality, addition and multiplication. The ease with which all these operations are performed is largely due to the admission of arbitrary local integers as coefficients.

7. $\varkappa(\mathfrak{p})$ and $\varkappa(\mathfrak{P})$

From k we pass to a finite field $\varkappa$ over k of degree n which we now assume once and for all to be non-degenerate. It would be most reasonable to call integral at $\mathfrak{p}$ a number γ in $\varkappa$ if it satisfies an equation

$$\gamma^r + c_1\gamma^{r-1} + \ldots + c_r = 0$$

whose coefficients c are numbers in k, locally integral at $\mathfrak{p}$. One readily sees that this definition is equivalent to the other: γ is locally integral at $\mathfrak{p}$ if it may be written as a fraction α/b whose numerator α is an integer in $\varkappa$ while the denominator b is an integer in k indivisible by $\mathfrak{p}$.

Let $\omega_1, \ldots, \omega_n$ be a basis for $\varkappa/k$. One can ascertain an integer c in k such that

$$c\omega_1 = \omega_1^*, \ldots, c\omega_n = \omega_n^*$$

are integers. Assume the discriminant of the basis $\omega_1^*, \ldots, \omega_n^*$ to be exactly divisible by the hth power of $\mathfrak{p}$.

<u>Lemma III 7, A</u>. *If α is locally integral at $\mathfrak{p}$, its components $a^{(i)}$ with respect to the basis ω_i,*

$$\alpha = \sum_i a^{(i)} \omega_i \qquad (a^{(i)} \text{ in } k)$$

are of order $\geq -h$ at $\mathfrak{p}$.

<u>Proof</u>. The equation

$$\alpha = \sum_i a_*^{(i)} \omega_i^* \qquad (ca_*^{(i)} = a^{(i)})$$

leads to

$$S(\alpha\omega_i^*) = \sum_k S(\omega_i^*\omega_k^*) \cdot a_*^{(k)}.$$

In solving them with respect to $a_*^{(i)}$ we see that $a_*^{(i)}$ and hence a fortiori a_i are of the type described in the lemma.

$\mathfrak{p}$-adic numbers in $\varkappa$ are introduced as formal series

$$\Gamma = \sum_{\nu=-\infty}^{+\infty} \gamma_\nu p^\nu \tag{7.1}$$

with coefficients γ_ν which are locally integral at $\mathfrak{p}$ and whose sequence breaks off towards $\nu \to -\infty$. It is evident how to add and multiply them. They form a commutative ring $\varkappa(\mathfrak{p})$.

<u>Theorem III 7, B.</u> *The ring $\varkappa(\mathfrak{p})$ is of degree n with respect to $k(\mathfrak{p})$.*

<u>Proof</u>. Relative to the basis $\omega_1, \ldots, \omega_n$ of $\varkappa/k$ we set

$$\gamma_\nu = \frac{c_\nu^{(1)}\omega_1 + \ldots + c_\nu^{(n)}\omega_n}{p^h} .$$

The $c_\nu^{(i)}$ are local integers in k at $\mathfrak{p}$ and we find for (7.1) the expression

$$C^{(1)}\omega_1 + \ldots + C^{(n)}\omega_n$$

with the $\mathfrak{p}$-adic coefficients in k:

$$C^{(i)} = \sum_\nu c_\nu^{(i)} p^{\nu-h}$$

This lagging behind of the exponent ν by h settles a point about which we were a little hazy at the end of paragraph (III) of the previous section.

We now come to the main point of the theory:

<u>Theorem III 7, C.</u> *$\varkappa(\mathfrak{p})$ breaks up into the direct sum of the fields $\varkappa(\mathfrak{P}_1), \ldots, \varkappa(\mathfrak{P}_g)$, parallel with the prime factorization of $\mathfrak{p}$ in $\varkappa$:*

$$\mathfrak{p} = \mathfrak{P}_1^{e_1} \cdots \mathfrak{P}_g^{e_g}$$

($\mathfrak{P}_1, \ldots, \mathfrak{P}_g$ distinct prime ideals in $\varkappa$; $e_1, \ldots, e_g > 0$).

We choose prime numbers $\pi_1, \ldots, \pi_g$ to $\mathfrak{P}_1, \ldots, \mathfrak{P}_g$. A number Γ in $\varkappa(\mathfrak{p})$ is at the same time a number in $\varkappa(\mathfrak{P})$ if $\mathfrak{P}$ is one of the g prime ideals $\mathfrak{P}_1, \ldots, \mathfrak{P}_g$:

$$\sum_\nu \alpha_\nu p^\nu = \sum_\nu \alpha_\nu^* \pi^{e\nu} \quad \text{with}$$
$$\alpha_\nu^* = \alpha_\nu \rho^\nu, \qquad p = \pi^e \cdot \rho.$$

α_ν which is integral at $\mathfrak{p}$ is a fortiori integral at $\mathfrak{P}$, and ρ is a unit at $\mathfrak{P}$. Hence Γ coincides with definite numbers $\Gamma^{(1)}, \ldots, \Gamma^{(g)}$ in $\varkappa(\mathfrak{P}_1), \ldots, \varkappa(\mathfrak{P}_g)$:

$$(7.2) \qquad \Gamma \to (\Gamma^{(1)}, \ldots, \Gamma^{(g)}).$$

Conversely if $\Gamma^{(1)}, \ldots, \Gamma^{(g)}$ are given numbers in $\varkappa(\mathfrak{P}_1), \ldots, \varkappa(\mathfrak{P}_g)$, there is a uniquely determined number Γ in $\varkappa(\mathfrak{p})$ such that the relation (7.2) holds.

<u>Proof</u>:

$$\Gamma^{(1)} = \sum_\nu \gamma_\nu^{(1)} \pi_1^\nu, \ldots, \Gamma^{(g)} = \sum_\nu \gamma_\nu^{(g)} \pi_g^\nu.$$

We introduce the partial sums like

$$\Gamma_\nu^{(1)} = \Sigma \gamma_i^{(1)} \pi_1^i \quad (i < e_1 \nu).$$

The sought-for number

$$\Gamma = \sum_\nu \gamma_\nu p^\nu$$

in $\varkappa(\mathfrak{p})$ will have the partial sums

$$\Gamma_\nu = \Sigma \gamma_i p^i \qquad (i < \nu).$$

One has to satisfy the simultaneous congruences

$$\Gamma_\nu \equiv \Gamma_\nu^{(1)} \ (\mathfrak{P}_1^{e_1 \nu}), \ldots, \Gamma_\nu \equiv \Gamma_\nu^{(g)} \ (\mathfrak{P}_g^{e_g \nu}).$$

Assume that one has succeeded in doing so for the exponent ν; we show how to proceed to $\nu + 1$. Set

$$(7.3) \qquad \Gamma_{\nu+1}^{(1)} = \Gamma_\nu^{(1)} + \bar{\gamma}_\nu^{(1)} \cdot p^\nu$$

$$\left\{ \bar{\gamma}_\nu^{(1)} = \left(\frac{\pi_1^{e_1}}{p}\right)^\nu \cdot \sum_{0 \leq i < e} \gamma_{e\nu+i}^{(1)} \cdot \pi_1^i \right\},$$

etc. One has to solve the simultaneous congruences

$$\gamma_\nu \equiv \bar{\gamma}_\nu^{(1)} \ (\mathfrak{P}_1^{e_1}), \ldots, \gamma_\nu \equiv \bar{\gamma}_\nu^{(g)} \ (\mathfrak{P}_g^{e_g}),$$

and we know that this is possible and determines γ_ν uniquely modulo

$$\mathfrak{P}_1^{e_1} \cdots \mathfrak{P}_g^{e_g} = \mathfrak{p}.$$

As so often in algebra, the invention of the right concepts is the chief burden; this accomplished, the proof of the main propositions becomes almost trivial.

Notice that Γ is integral in $\varkappa(\mathfrak{y})$ if and only if $\Gamma^{(1)},\ldots,\Gamma^{(g)}$ are integral in $\varkappa(\mathfrak{z}_1),\ldots,\varkappa(\mathfrak{z}_g)$.

We denote by $n_1,\ldots,n_g$ the degrees of the fields $\varkappa(\mathfrak{z}_1),\ldots,\varkappa(\mathfrak{z}_g)$ relative to $k(\mathfrak{y})$:

$$n = n_1 + \ldots + n_g.$$

A basis for $\varkappa(\mathfrak{y})$ is obtained from bases of the g fields $\varkappa(\mathfrak{z}_1),\ldots,\varkappa(\mathfrak{z}_g)$ in the following manner. Let $\Omega_1^{(1)},\ldots,\Omega_{n_1}^{(1)}$ be a basis of $\varkappa(\mathfrak{z}_1)$. We denote by $\Omega_i^{(1)}$ that number in $\varkappa(\mathfrak{y})$ which $= \Omega_i^{(1)}$ in $\varkappa(\mathfrak{z}_1)$, but $= 0$ in $\varkappa(\mathfrak{z}_2)$ and ... and in $\varkappa(\mathfrak{z}_g)$. Then

$$\Omega_1^{(1)},\ldots,\Omega_{n_1}^{(1)};\ \ldots;\ \Omega_1^{(g)},\ldots,\Omega_{n_g}^{(g)} \tag{7.4}$$

constitute a basis for $\varkappa(\mathfrak{y})$. The representing matrix of a number Γ of $\varkappa(\mathfrak{y})$ relative to this basis decomposes into g parts corresponding to the g partial fields. Therefore

$$\left\{\begin{aligned} S(\Gamma) &= S_1(\Gamma) + \ldots + S_g(\Gamma),\\ Nm(\Gamma) &= Nm_1(\Gamma)\ \ldots\ Nm_g(\Gamma);\\ f(t) &= f_1(t)\ \ldots\ f_g(t). \end{aligned}\right. \tag{7.5}$$

The traces and norms on the left side are taken in $\varkappa(\mathfrak{y})$, the partial traces and norms on the right side in the fields $\varkappa(\mathfrak{z}_1),\ldots,\varkappa(\mathfrak{z}_g)$ always relative to $k(\mathfrak{y})$. $f(x)$ is the field equation $Nm(x - \Gamma)$. If $\Gamma = \theta$ is a determining number of $\varkappa/k$ it is also a determining number in $\varkappa(\mathfrak{z}_1),\ldots,\varkappa(\mathfrak{z}_g)$, and the factors $f_1(x),\ldots,f_g(x)$ are irreducible in $k(\mathfrak{y})$. The discriminant of the basis (7.4) is the product of the partial discriminants of the bases

$$\begin{array}{lll} \Omega_1^{(1)},\ldots,\Omega_{n_1}^{(1)} & \text{in} & \varkappa(\mathfrak{z}_1),\\ \ldots & & \ldots\\ \Omega_1^{(g)},\ldots,\Omega_{n_g}^{(g)} & \text{in} & \varkappa(\mathfrak{z}_g). \end{array}$$

In this, as in all other questions, the study of $\varkappa(\mathfrak{y})$ is definitely reduced to that of the direct summands $\varkappa(\mathfrak{z}_1),\ldots,\varkappa(\mathfrak{z}_g)$.

We take one of these, $\varkappa(\mathfrak{z})$, call its degree m, and construct a special basis for it. Let $\varkappa_{\mathfrak{z}}$, the field of residues modulo $\mathfrak{P}$, be of degree f relative to $k_{\mathfrak{p}}$; f is called the (Kummer) degree of $\mathfrak{P}$. With respect to a basis $\tau_1, \ldots, \tau_f$ for $\varkappa_{\mathfrak{P}}/k_{\mathfrak{p}}$ we have for any number α in $\varkappa$ which is locally integral at $\mathfrak{P}$:

$$(7.6) \qquad \alpha \equiv a_1\tau_1 + \ldots + a_f\tau_f \pmod{\mathfrak{P}}.$$

The components $a_1, \ldots, a_f$ are integers in k, uniquely determined if chosen from a fixed system Σ of residues modulo $\mathfrak{p}$. First $\tau_1, \ldots, \tau_f$ are numbers of $\varkappa_{\mathfrak{P}}$ but we fix them as integers in $\varkappa$ (they will ultimately matter modulo $\mathfrak{P}^e$). From (7.6) one derives a congruence

$$\alpha \equiv \Sigma a_{ik}\tau_i\pi^k \pmod{\mathfrak{P}^e}$$

$$(i = 1, \ldots, f; \; k = 0, 1, \ldots, e - 1)$$

with uniquely determined coefficients a_{ik} in Σ.

Write any $\mathfrak{P}$-adic number in the form

$$\sum_\nu \overline{\gamma}_\nu p^\nu$$

which arises from the standard form $\Sigma\gamma_\nu\pi^\nu$ by summing in leaps of e terms; cf. (7.3). Then one realizes that the ef numbers

$$(7.7) \qquad \tau_i\pi^k \quad (i = 1, \ldots, f; \; k = 0, 1, \ldots, e - 1)$$

which we denote in any fixed arrangement by $\omega_1, \ldots, \omega_m$, constitute a basis for $\varkappa(\mathfrak{P})/k(\mathfrak{p})$, and more precisely an integral basis; i.e., if Γ is an integer in $\varkappa(\mathfrak{P})$, then the components C_i of Γ relative to this basis,

$$\Gamma = C_1\omega_1 + \ldots + C_m\omega_m,$$

are integers in $k(\mathfrak{p})$. The degree $m = ef$.

In this way we find

$$n_1 = e_1f_1, \ldots, n_g = e_gf_g$$

and the relation

$$(7.8) \qquad e_1f_1 + \ldots + e_gf_g = n.$$

By putting together the bases for $\varkappa(\mathfrak{P}_1), \ldots, \varkappa(\mathfrak{P}_g)$ in the manner described above we arrive at

Theorem III 7, D. *$\varkappa(\mathfrak{p})$ has an integral basis $\Omega_1, \ldots, \Omega_n$ over $k(\mathfrak{p})$.*

If one carries the construction through modulo $\mathfrak{P}_1^{e_1}, \ldots, \mathfrak{P}_g^{e_g}$ rather than in the realms of $\mathfrak{P}_1, \ldots, \mathfrak{P}_g$, one finds such an integral basis which consists of ordinary numbers $\omega_1, \ldots, \omega_n$ in $\varkappa$. One may derive the same result from the last theorem by limiting oneself to the initial terms ω_i in the expansions

$$\Omega_i = \omega_i + \omega_i' p + \ldots \qquad \text{in } \varkappa(\mathfrak{p}).$$

Indeed, the initial terms of the equation

$$\gamma = C_1\Omega_1 + \ldots + C_n\Omega_n$$

yield

$$\gamma \equiv c_1\omega_1 + \ldots + c_n\omega_n \quad (\text{mod } \mathfrak{p})$$

for a local integer γ at $\mathfrak{p}$ in $\varkappa$; c_i integers in k. This proves $\omega_1, \ldots, \omega_n$ to be a basis for $\varkappa(\mathfrak{p})/k(\mathfrak{p})$: any integral number Γ in $\varkappa(\mathfrak{p})$ is representable as a sum

$$C_1\omega_1 + \ldots + C_n\omega_n$$

with integral components C_i in $k(\mathfrak{p})$. If $\Gamma = \gamma$ is an ordinary number in $\varkappa$ which is integral at $\mathfrak{p}$, then the coefficients C_i will turn out as ordinary numbers in k as follows from the equations

$$S(\gamma\omega_i) = \sum_k S(\omega_i\omega_k) \cdot C_k$$

for C_i with a non-vanishing determinant.

Theorem III 7, E. *$\varkappa$ has a basis $\omega_1, \ldots, \omega_n$ over k which is locally integral at $\mathfrak{p}$.*

This means: $\omega_1, \ldots, \omega_n$ are local integers, and if γ in $\varkappa$ is integral at $\mathfrak{p}$, then its components with respect to this basis are locally integral at $\mathfrak{p}$ in k. This result is only appreciated fully when one observes that in general a finite field $\varkappa/k$ is far from having an integral basis in the absolute sense.

In conclusion we show:

Theorem III 7, F. *The Kummer degree f of a prime ideal in $\varkappa$ equals its Kronecker degree.*

Proof. We compute the norm of the prime number π to $\mathfrak{P}$ by means of the basis (7.7). Notice that

$$\pi^{e} = p \cdot \rho \tag{7.9}$$

where ρ is a unit at $\mathfrak{P}$. We have congruences

$$\rho\tau_i \equiv \sum_k r_{ik}\tau_k \qquad (\text{mod } \mathfrak{P})$$

with integral coefficients r_{ik} in k (or in $k_{\mathfrak{p}}$). By definition $|r_{ik}|$ is the Nm (ρ) in $\varkappa_{\mathfrak{P}}$ and hence a unit in $k_{\mathfrak{p}}$. (Indeed, $\rho\rho' = 1$, with ρ and ρ' being integral at $\mathfrak{P}$, leads to

$$\text{Nm } \rho \cdot \text{Nm } \rho' = 1 \quad \text{in } \varkappa_{\mathfrak{P}}.)$$

Considering that

$$\pi \cdot \tau_i\pi^{k-1} = \tau_i\pi^{k} \qquad (k = 1, \ldots, e - 1)$$

$$\pi \cdot \tau_i\pi^{e-1} = p \cdot \tau_i\rho$$

one finds the first f(e - 1) lines of the representing matrix of π in $\varkappa(\mathfrak{P})$ to be of the form

$$\begin{Vmatrix} 0 & E & 0 & \ldots & 0 \\ 0 & 0 & E & \ldots & 0 \\ \cdot & \cdot & \cdot & \cdot & \cdot \\ 0 & 0 & 0 & \ldots & E \end{Vmatrix}$$

where E is the f-dimensional unit matrix, while the last f lines are divisible by p and in the lower left corner make up the matrix

$$p(r_{ik} + r'_{ik}p + \ldots) \qquad [i, k = 1, \ldots, f].$$

Consequently the norm

$$\text{Nm } \pi \quad \text{of } \pi \quad \text{in } \pi(\mathfrak{P})/k(\mathfrak{p})$$

is exactly divisible by $\mathfrak{p}^{f}$.

If ε is a unit at $\mathfrak{P}$, the norm of ε in $\varkappa(\mathfrak{P})/k(\mathfrak{p})$ is a unit at $\mathfrak{p}$ in k (see the above remark about ρ).

Let $\pi = \pi_1$ be divisible by $\mathfrak{P} = \mathfrak{P}_1$ but such that $(\pi)/\mathfrak{P}$ is prime to $\mathfrak{p}$. Then π is a unit in $\varkappa(\mathfrak{P}_2), \ldots, \varkappa(\mathfrak{P}_g)$, and the formula (7.5) shows that the total norm

$$\mathrm{Nm}\,\pi \quad \text{of } \pi \quad \text{in } \varkappa/k$$

is exactly divisible by $\mathfrak{p}^f$. Comparison with the criterion for the Kronecker degree f, as given in Theorem II 8, G and its proof, reveals the identity of both degrees.

8. Discriminant

An important new subject is taken up in this section in which we shall be led to ascribe a discriminant to the field itself rather than to a special basis of the field.

Let us suppose first that $\varkappa/k$ has an integral basis $\omega_1, \ldots, \omega_n$ and call its discriminant d. An arbitrary integral basis $\omega_1^*, \ldots, \omega_n^*$ will be connected with the basis ω_i by equations

$$\omega_i^* = \sum_k a_{ik}\omega_k, \qquad \omega_i = \sum_k b_{ik}\omega_k^*$$

where a_{ik}, b_{ik} are integers in k. Hence for their discriminants d, d^* the equations

$$d^* = |a_{ik}|^2 \cdot d, \qquad d = |b_{ik}|^2 \cdot d^*$$

hold, which show that d and d^* are associate. Thus the principal ideal $\vartheta = (d)$ which we call the discriminant of the field is fixed unambiguously.

We next turn to an arbitrary non-degenerate field $\varkappa$ of finite degree n over a Dedekind ground field k, and apply the above remark to

$$\varkappa(\mathfrak{P})/k(\mathfrak{p}) \quad \text{and} \quad \varkappa(\mathfrak{p})/k(\mathfrak{p}).$$

The first field has an integral basis and its discriminant D is an integral number of $k(\mathfrak{p})$. The discriminant D^* of any integral basis is $\sim D$, and hence the order t of D at $\mathfrak{p}$,

$$D \sim \mathfrak{p}^t,$$

is independent of the choice of the integral basis. Out of integral bases for $\varkappa(\mathfrak{P}_1), \ldots, \varkappa(\mathfrak{P}_g)$ with the discriminants

$$D_1 \sim \mathfrak{p}^{t_1}, \ldots, D_g \sim \mathfrak{p}^{t_g}$$

we have combined an integral basis for $\kappa(\mathfrak{p})/k(\mathfrak{p})$, and its discriminant

$$D = D_1 \ldots D_g \sim \mathfrak{p}^t \quad \text{with} \quad t = t_1 + \ldots + t_g.$$

The product

$$\vartheta = \prod_{\mathfrak{p}} \mathfrak{p}^t \tag{8.1}$$

extending to all prime ideals $\mathfrak{p}$ of k is called the <u>discriminant of</u> k. The difference, as compared with the first case, arises from the necessity to pick for each individual prime divisor $\mathfrak{p}$ its contribution to the discriminant because we have at our disposal a <u>local</u> integral basis at each $\mathfrak{p}$, but not a universal integral basis. For the same reason one must not expect ϑ to be a principal ideal.

The product (8.1) has a meaning since the exponent $t = 0$ for almost all $\mathfrak{p}$, that is to say with only a finite number of exceptions. Indeed, if $\omega_1, \ldots, \omega_n$ is a basis of κ/k which consists of integers, one readily sees that its discriminant $d(\omega_1, \ldots, \omega_n)$ is divisible at least by the tth power of $\mathfrak{p}$, and ϑ is therefore a divisor of $d(\omega_1, \ldots, \omega_n)$.

A closer investigation of the discriminant will be carried out under the assumption that for each pair $\mathfrak{p} : \mathfrak{P}$, i.e., for any prime ideal $\mathfrak{p}$ in k and any prime divisor $\mathfrak{P}$ of $\mathfrak{p}$ in κ, the residue field $\kappa_{\mathfrak{P}}$ is non-degenerate over $k_{\mathfrak{p}}$. We shall then say κ/k is locally non-degenerate everywhere. This hypothesis guarantees the existence of a determining number τ_o of $\kappa_{\mathfrak{P}}/k_{\mathfrak{p}}$. It will satisfy an irreducible equation

$$g(\tau_o) = \tau_o^f + b_1\tau_o^{f-1} + \ldots + b_f = 0$$

in $k_{\mathfrak{p}}$. We fix the coefficients b_i which count here merely mod $\mathfrak{p}$ in a definite manner as integers in k. The differential $\dot{g}(\tau_o)$ is indivisible by $\mathfrak{P}$. This enables one to play a curious trick, namely to ascertain a $\mathfrak{P}$-adic number

$$\tau = \tau_o + \tau_1\pi + \tau_2\pi^2 + \ldots$$

which in $\kappa(\mathfrak{P})$ satisfies the <u>equation</u>

$$g(\tau) = 0.$$

In fact, after one has found

$$\sigma_\nu = \tau_o + \tau_1\pi + \ldots + \tau_{\nu-1}\pi^{\nu-1}$$

such that

$$g(\sigma_\nu) \equiv 0 \qquad (\text{mod } \mathfrak{P}^\nu)$$

one solves the congruence

$$(8.2) \qquad g(\sigma_{\nu+1}) \equiv 0 \qquad (\mathfrak{P}^{\nu+1})$$

by

$$\sigma_{\nu+1} = \sigma_\nu + \tau_\nu \pi^\nu.$$

As for $\nu \geqq 1$:

$$g(\sigma_\nu + \tau_\nu\pi^\nu) \equiv g(\sigma_\nu) + \dot{g}(\sigma_\nu)\tau_\nu\pi^\nu$$
$$\equiv g(\sigma_\nu) + \dot{g}(\tau_o)\tau_\nu\pi^\nu \quad (\mathfrak{P}^{\nu+1}),$$

(8.2) reduces to

$$\dot{g}(\tau_o) \cdot \tau_\nu \equiv -\frac{g(\sigma_\nu)}{\pi^\nu} \quad (\text{mod } \mathfrak{P}),$$

and the latter congruence is solvable because $\dot{g}(\tau_o)$ is a unit at $\mathfrak{P}$.

The numbers

$$(8.3) \quad \tau^i\pi^k \quad (i = 0, \ldots, f - 1; \ k = 0, \ldots, e - 1)$$

constitute an integral basis of $\varkappa(\mathfrak{P})/k(\mathfrak{p})$. By adjunction of τ the field $k(\mathfrak{p})$ changes into a field $\overline{k}$ of degree f over $k(\mathfrak{p})$, and $\varkappa(\mathfrak{P})$ is a field of relative degree e over the intermediary field $\overline{k}$, with π as determining number. Every integer in $\varkappa(\mathfrak{P})$ may be uniquely written as

$$\overline{c}_o + \overline{c}_1\pi + \ldots + \overline{c}_{e-1}\pi^{e-1}$$

with coefficients $\overline{c}$ which are integers in $\overline{k}$:

$$\overline{c} = c_o + c_1\tau + \ldots + c_{f-1}\tau^{f-1}$$

(c_i integers in $k(\mathfrak{p})$). If we apply this to the number ρ in (7.9) we see that the irreducible equation which π satisfies relative to $\overline{k}$ is of the form

$$f(\pi) = \pi^e + p(\overline{a}_1\pi^{e-1} + \ldots + \overline{a}_{e-1}) = 0$$

where $\overline{a}_{e-1}$ is a unit at $\mathfrak{P}$.

The general Theorem I 4, B, about discriminants is applicable to the tower

$$k(\mathfrak{y}) \subset \bar{k} \subset \varkappa(\mathfrak{z})$$

and the telescopic basis (8.3). The small discriminant of the basis

$$1, \tau, \ldots, \tau^{f-1}$$

is a unit at $\mathfrak{z}$ (because $\varkappa_{\mathfrak{z}}/k_{\mathfrak{y}}$ is non-degenerate). The relative discriminant of the basis

$$1, \pi, \ldots, \pi^{e-1}$$

is the relative norm of $\dot{f}(\pi)$. Hence the large discriminant is associate to the total norm of $\dot{f}(\pi)$. Let

$$\dot{f}(\pi) = e\pi^{e-1} + p\left\{(e-1)\bar{a}_1\pi^{e-2} + \ldots\right\}$$

be exactly divisible by the power $\varepsilon - 1$ of $\mathfrak{z}$. One has $\varepsilon = e$ or $\varepsilon > e$ according as $e1$ is not or is divisible by $\mathfrak{z}$.

After composing $\varkappa(\mathfrak{y})$ out of the g fields $\varkappa(\mathfrak{z}_1), \ldots, \varkappa(\mathfrak{z}_g)$ by direct summation, one finds:

Theorem III 8, A. *Under the hypothesis that $\varkappa/k$ is locally non-degenerate everywhere, the discriminant of $\varkappa$ is the norm of a certain ideal in $\varkappa$, the ramification ideal*

$$\vartheta = \prod_{\mathfrak{z}} \mathfrak{z}^{\varepsilon-1}.$$

The exponent $\varepsilon - 1$ is determined by

$$f(\pi) \sim \mathfrak{z}^{\varepsilon-1}$$

If $\mathfrak{y}$ is exactly divisible by the eth power of $\mathfrak{z}$, one has in general $\varepsilon = e$, but $\varepsilon > e$ if $e1 : \mathfrak{z}$.

The explicit formula for the discriminant is

$$\vartheta = \prod_{\mathfrak{y}} \mathfrak{y}^{(\varepsilon_1-1)f_1 + \ldots + (\varepsilon_g-1)f_g} \tag{8.4}$$

Hence the important

Corollary *(Dedekind): $\mathfrak{p}$ goes into the discriminant of κ/k if and only if $\mathfrak{p}$ contains multiple prime ideals in κ.*

The hypothesis of local non-degeneracy is certainly fulfilled if k is a numerical field; for then $\kappa_{\mathfrak{P}}$ is a strictly finite field over $k_{\mathfrak{p}}$ and therefore non-degenerate (and even Galois with a cyclic group of order f). el : $\mathfrak{P}$ in this case means divisibility of the exponent e by the prime characteristic of $k_{\mathfrak{p}}$.

9. Relative Discriminant

We consider our familiar tower $k \subset \kappa \subset K$ erected over a Dedekind field k and with two non-degenerate stories κ/k and K/κ. Let $\mathfrak{p}_0$, $\mathfrak{p}$, $\mathfrak{P}$ be prime ideals in k, κ, K respectively such that $\mathfrak{p}_0 : \mathfrak{p} : \mathfrak{P}$ and let $\mathfrak{p}^{e_0}$, $\mathfrak{P}^e$ be the exact powers going into $\mathfrak{p}_0$ and $\mathfrak{p}$ respectively. Then $\mathfrak{p}_0$ is exactly divisible by $\mathfrak{P}^E$ where

$$E = e_0 e$$

(great exponent = small exponent x relative exponent). If f_0, f, F are the degree of $\mathfrak{p}$, relative and absolute degrees of $\mathfrak{P}$, then

$$F = f_0 f,$$

the reason being that f_0, f, F are small degree, relative degree and large degree in the two-story tower

$$k_{\mathfrak{p}_0} \subset \kappa_{\mathfrak{p}} \subset K_{\mathfrak{P}}.$$

To these almost trivial instances of multiplicative behavior one can add the ramification ideal; with

$$\vartheta(\kappa), \quad \vartheta(K/\kappa), \quad \vartheta(K)$$

denoting the ramification ideals of the fields indicated as arguments, one has

$$\vartheta(K) = \vartheta(\kappa) \cdot \vartheta(K/\kappa) \tag{9.1}$$

if κ and K/κ are locally non-degenerate. We summarize:

Theorem III 9, A. *In a tower of relative fields over a Dedekind ground field k, exponents and degrees of prime ideals satisfy the multiplicative law.*

Theorem III 9, B. *In a tower of relative, locally non-degenerate fields, the ramification ideals satisfy the multiplicative law.*

The proof for the ramification ideals depends on a formula which we shall develop first. Let $\varkappa = k(\theta)$ be a simple extension of degree n over the ground field k, and $K = \varkappa(\Theta)$ a simple extension of degree r over $\varkappa$, with the determining equations

$$(9.2) \qquad f(x) = 0 \quad \text{for } \theta \quad \text{and} \quad \varphi(x) = 0 \quad \text{for } \Theta$$

respectively. We assume that

$$1, \theta, \ldots, \theta^{n-1} \quad \text{and} \quad 1, \Theta, \ldots, \Theta^{r-1}$$

are integral bases for $\varkappa$ and $K/\varkappa$ respectively. The coefficients of the two equations (9.2) are then integers in k and $\varkappa$ respectively. The field equation for Θ in k is of degree $N = nr$,

$$(9.3) \qquad F(x) = Nm_{\varkappa}\varphi(x) = \varphi(x) \cdot \lambda(x)$$

where the "tail" $\lambda(x)$ also has integral coefficients. We propose to establish the following equation for the differentials

$$(9.4) \qquad \dot{F}(\Theta) = \dot{f}(\theta)\ \dot{\varphi}(\Theta)\Lambda$$

with Λ an integer in K.

By deriving (9.3), one gets

$$\dot{F}(x) \equiv \dot{\varphi}(x) \cdot \lambda(x) \qquad \{\text{mod. } \varphi(x)\}$$

Put

$$f(x) = Nm_{\varkappa}(x - \theta) = (x - \theta) \cdot \psi(x).$$

We now operate in the universe U embedding the field $\varkappa$; f(x) decomposes in U into n conjugate linear factors:

$$\psi(x) = (x - \theta') \ldots (x - \theta^{(n-1)}).$$

$$\lambda(x) = \varphi'(x) \ldots \varphi^{(n-1)}(x).$$

[It would be artificial to introduce an embedding field which includes Θ besides the conjugates $\theta, \theta', \ldots$.] We express the coefficients of $\varphi(x)$ by means of the basis $1, \theta, \ldots, \theta^{n-1}$,

$$\varphi(x) = g(x,\theta)$$

where $g(x,y)$ is a polynomial of two variables in k with integral coefficients, and then introduce a polynomial $G(x;y,y^*)$ in k of three variables x; y, y^* by

$$g(x,y) - g(x,y^*) = (y - y^*) \cdot G(x;y,y^*).$$

We set

$$\mathrm{Nm}_{\varkappa}\, G(x;y,\theta) = G(x;y,\theta) \cdot \eta^*(x,y).$$

η^* is a polynomial in $\varkappa$ with integral coefficients and equals the product

$$G(x,y,\theta') \ldots G(x;y,\theta^{(n-1)}).$$

Next we substitute $y = \theta$ and with

$$\eta^*(x,\theta) = \eta(x)$$

find the equation

$$G(x;\theta,\theta') \ldots G(x;\theta,\theta^{(n-1)}) = \eta(x)$$

or

$$\left\{ g(x,\theta) - g(x,\theta') \right\} \ \ldots \ \left\{ g(x,\theta) - g(x,\theta^{(n-1)}) \right\}$$
$$= (\theta - \theta') \ldots (\theta - \theta^{(n-1)}) \cdot \eta(x).$$

Consequently

$$\lambda(x) = g(x,\theta') \ldots g(x,\theta^{(n-1)})$$
$$\equiv (-1)^{n-1}\dot{f}(\theta)\eta(x) \qquad \left\{ \mathrm{mod.}\ \varphi(x) \right\},$$

and this results in (9.4) with

$$\Lambda = (-1)^{n-1}\eta(\theta).$$

We now make the further assumption that

(9.5) $$1, \Theta, \ldots, \Theta^{N-1}$$

is an integral basis for K/k, and then show that Λ is a unit,

$$\dot{F}(\Theta) \sim \dot{f}(\theta) \cdot \dot{\varphi}(\Theta),$$

or that the differential $\dot{F}(\Theta)$ of K is associate with the product of the differential $\dot{f}(\theta)$ of $\varkappa$ by the relative differential $\dot{\varphi}(\Theta)$ of $K/\varkappa$. Taking the absolute norm of (9.4) one gets

$$D = d^r \cdot Nm_{\varkappa}\, \delta \cdot Nm_K\, \Lambda$$

where

$$D = d(1, \Theta, \ldots, \Theta^{N-1}), \qquad d = D(1, \theta, \ldots, \theta^{n-1})$$

are the discriminants of K and $\varkappa$ with respect to the indicated bases, while δ is the relative discriminant

$$\delta = D(1, \ldots, \Theta^{r-1})$$

in $K/\varkappa$. On the other hand, by Theorem I 4, B, the discriminant D^* of the telescopic basis

(9.6) $$\theta^i\Theta^k \quad (i = 0, \ldots, n-1;\ k = 0, \ldots, r-1)$$

for K satisfies the equation

$$D^* = d^r \cdot Nm_{\varkappa}\, \delta.$$

Therefore

$$D = D^* \cdot Nm\, \Lambda.$$

But if (9.5) is an integral basis as well as (9.6), then D differs from D^* by a factor in k which is a unit; consequently

$$Nm\, \Lambda \sim 1 \text{ in } k \text{ and thus } \Lambda \sim 1 \text{ in } K.$$

After these preliminary considerations we return to the situation referred to in Theorem III 9, B. $K(\mathfrak{P})$ is reached from $k(\mathfrak{y}_0)$ by two consecutive adjunctions, T of degree F and Π of degree E,

$$k(\mathfrak{y}_0) \subset \bar{\bar{k}} \subset K(\mathfrak{P}).$$

T and Π have the same significance for $K(\mathfrak{P})$ as τ and π for $\varkappa(\mathfrak{y})$. However we must be a little more explicit about T. The primitive residue T_o mod $\mathfrak{P}$ satisfies a congruence $\gamma_o(x) \equiv 0(\mathfrak{P})$ of degree f with coefficients $\bar{g}_1(\tau_o),\ldots,\bar{g}_f(\tau_o)$ in $\varkappa_{\mathfrak{y}}$. The $\bar{g}_i$ are polynomials in the ground field k. We replace the coefficients $\bar{g}_i(\tau_o)$ by $\bar{g}_i(\tau)$, thus obtaining a congruence

$$\gamma(T_o) \equiv 0 \quad (\mathfrak{P})$$

with coefficients in $\bar{k} = k(\mathfrak{y}_o|\tau)$, and as described before construct the number T in $K(\mathfrak{P})$ which is $\equiv T_o(\mathfrak{P})$ and satisfies the equation $\gamma(T) = 0$. This has the effect that $k(\mathfrak{y}_o|\tau) = \bar{k}$ is a subfield of $k(\mathfrak{y}_o|T) = \bar{\bar{k}}$. The $\mathfrak{P}$-adic developments of numbers in $K(\mathfrak{P})$ show that

$$1, \Pi, \ldots, \Pi^{E-1}$$

as well as

$$\Pi^i\pi^j \quad (i = 0,\ldots, e - 1;\ j = 0,\ldots, e_o - 1)$$

constitute an integral basis for $K(\mathfrak{P})/\bar{\bar{k}}$. As π satisfies an equation of degree e_o in $\bar{k}$ and thus in $\bar{\bar{k}}$, the tower

$$\bar{\bar{k}} \subset \bar{\bar{k}}(\pi) \subset K(\mathfrak{P})$$

satisfies the assumptions on which the preliminary argument has been based with

$$e_o,\ e,\ \pi,\ \Pi \quad \text{instead of} \quad n,\ r,\ \theta,\ \Theta,$$

and we thus convince ourselves that any prime ideal $\mathfrak{P}$ in K contributes the same factor to both sides of the equation (9.1).

By forming the norm of that equation, we find for the discriminants

$$\vartheta_o(\varkappa), \qquad \vartheta(K/\varkappa) \qquad \text{and} \qquad \vartheta_o(K)$$

(they are ideals in k, $\varkappa$ and k respectively), the following law holding under the same condition of local non-degeneracy:

Theorem III 9, C.

$$\vartheta_0(K) = \left\{\vartheta_0(\varkappa)\right\}^r \cdot Nm_\varkappa\, \vartheta(K/\varkappa).$$

The algebraic results obtained for discriminants of bases in Chapter I have thus been paralleled by corresponding arithmetical results for the discriminants of the fields themselves.

10. Hilbert's Theory of Galois Fields, Artin Symbol

In this section we assume $\varkappa = k(\theta)$ to be a Galois field of degree n over the Dedekind ground field k, and $\mathfrak{G} = \{s\}$ to be its Galois group. Following Hilbert, we analyze the decomposition of $\mathfrak{p}$ in passing from k to $\varkappa$ through various intermediary fields. As always $\mathfrak{p} : \mathfrak{P}$ denotes a pair of prime ideals in k and $\varkappa$. The substitutions s satisfying the condition $\mathfrak{P}^s = \mathfrak{P}$ form a subgroup γ of $\mathfrak{G}$, called splitting group, whose index may be designated by g. The substitutions of each of the cosets carry $\mathfrak{P}$ into the same conjugate $\mathfrak{P}'$. We thus obtain

$$\mathfrak{p}^f = Nm\,\mathfrak{P} = \prod_s \mathfrak{P}^s = (\mathfrak{P}_1 \cdots \mathfrak{P}_g)^{n/g}$$

where $\mathfrak{P}_1, \ldots, \mathfrak{P}_g$ are the distinct prime divisors of $\mathfrak{p}$ in $\varkappa$. Hence

$$\mathfrak{p} = (\mathfrak{P}_1 \cdots \mathfrak{P}_g)^e, \qquad n = efg.$$

The order of the splitting group γ is ef.

In the realm of $\mathfrak{p}$ the determining equation F(x) of θ decomposes into g irreducible factors of degree ef,

$$F(x) = F_1(x) \cdots F_g(x) \quad (\mathfrak{p}),$$

which correspond to the ideals $\mathfrak{P}_1, \ldots, \mathfrak{P}_g$. Consider the factor F_1 corresponding to $\mathfrak{P}_1 = \mathfrak{P}$. Since $\mathfrak{P}^s = \mathfrak{P}$, the equation $F_1(\theta) = 0$ holding in $\varkappa(\mathfrak{P})$, implies $F_1(\theta^s) = 0$ for every substitution s of the splitting group. Hence the simple result:

Theorem III 10, A. *$\varkappa(\mathfrak{P})$ is a Galois field over $k(\mathfrak{p})$ of degree ef, and Γ is its Galois group.*

Again we introduce the assumption that $\varkappa/k$ is locally non-degenerate everywhere, and the intermediary field $\bar{k} = k(\mathfrak{p} | \tau)$ with the determining equation

$$g(\tau) = \tau^f + \dots = 0.$$

$\bar{k}$ belongs to a certain subgroup $\mathfrak{T}$ of Γ which is called the inertial group. $\mathfrak{T}$ is the Galois group of $\varkappa(\mathfrak{P})/\bar{k}$ and its substitutions t satisfy the equation $\tau^t = \tau$. Two conjugates $\tau^s, \tau^{s'}$ of τ coincide if and only if s and s' (in Γ) are in the same coset mod $\mathfrak{T}$, that is to say, if s' = ts. Because $\varkappa_{\mathfrak{P}}$ is separable over $k_{\mathfrak{p}}$, τ is not only different from the f - 1 conjugates $\tau^{s_1}, \dots, \tau^{s_{f-1}}$ which correspond to the f - 1 cosets $\neq \mathfrak{T}$, but also incongruent mod $\mathfrak{P}$ to each of them.

Any integer α (or local integer at $\mathfrak{P}$) in $\varkappa$ satisfies the congruence

$$(10.1) \qquad \alpha^t \equiv \alpha \quad (\mathfrak{P}) \qquad \{t \text{ in } \mathfrak{T}\}.$$

Indeed, any integer in $\varkappa(\mathfrak{P})$ is representable in the form

$$(10.2) \qquad \rho_0 + \rho_1 \pi + \dots$$

where $\rho_0, \rho_1, \dots$ are numbers in $\bar{k}$ of the form

$$c_0 + c_1 \tau + \dots + c_{f-1} \tau^{f-1}$$

(c integers in k, chosen from a set Σ of residues mod $\mathfrak{p}$). Vice versa, a substitution t in Γ satisfying the congruence (10.1) for all integers α in $\varkappa$ lies in the inertial group. In fact, if t is not in $\mathfrak{T}$ then

$$\tau^t \not\equiv \tau \ (\mathfrak{P}) \quad \text{and hence} \quad \tau_0^t \not\equiv \tau_0 \ (\mathfrak{P})$$

with τ_0 denoting the residue in Σ which is $\equiv \tau$ ($\mathfrak{P}$). Let t be in $\mathfrak{T}$ and s in Γ. On account of $\mathfrak{P}^s = \mathfrak{P}$, the congruence (10.1) entails

$$\alpha^{ts} \equiv \alpha^s \ (\mathfrak{P}) \quad \text{or} \quad \beta^{s^{-1}ts} \equiv \beta \ (\mathfrak{P}) \quad \text{with } \beta = \alpha^s,$$

or $s^{-1}ts$ is in $\mathfrak{T}$. In other words, $\mathfrak{T}$ is an invariant subgroup of Γ, and hence $\bar{k}$ is a Galois field over $k(\mathfrak{p})$ with the factor group $\mathfrak{T}/\mathfrak{Z}$ as its Galois group. This means that $\varkappa_{\mathfrak{z}}/k_{\mathfrak{g}}$ is a Galois field and its automorphisms are the substitutions of $\mathfrak{T}$ mod $\mathfrak{Z}$. More explicitly, for any s in $\mathfrak{T}$ the congruence

$$\alpha' \equiv \alpha^s \ (\mathfrak{P}) \qquad \{\alpha \text{ any integer in } \varkappa\}$$

defines an automorphism $\alpha \longrightarrow \alpha'$ of $\varkappa_{\mathfrak{z}}/k_{\mathfrak{g}}$, and two s give rise to the same automorphism of $\varkappa_{\mathfrak{z}}/k_{\mathfrak{g}}$ if and only if they coincide in $\mathfrak{T}/\mathfrak{Z}$. For the special case where k is numerical, we have seen, independently of these considerations, that $\varkappa_{\mathfrak{P}}/k_{\mathfrak{g}}$ is a Galois field with a cyclic group generated by the automorphism

$$\alpha \rightarrow \alpha^P$$

where $P = N\mathfrak{g}$, a power of the prime characteristic of $k_{\mathfrak{g}}$, is the number of elements in $k_{\mathfrak{g}}$. Consequently the substitutions s of $\mathfrak{T}/\mathfrak{Z}$ may be arranged in such fashion $s_0 = 1, s_1, \ldots, s_{f-1}$ that

$$\alpha^{s_0} \equiv \alpha, \quad \alpha^{s_1} \equiv \alpha^P, \ldots, \alpha^{s_{f-1}} \equiv \alpha^{P^{f-1}} \quad (\text{mod } \mathfrak{P}).$$

We summarize:

Theorem III 10, B. *The elements t of $\mathfrak{T}$ which satisfy the congruence*

$$\alpha^t \equiv \alpha \ (\mathfrak{P})$$

for every integer α in $\varkappa$ form an invariant subgroup $\mathfrak{Z}$ of $\mathfrak{T}$. $\varkappa_{\mathfrak{z}}/k_{\mathfrak{g}}$ is a Galois field with the factor group $\mathfrak{T}/\mathfrak{Z}$ as its Galois group.

The subgroups $\mathfrak{T}$ and $\mathfrak{Z}$ determine subfields k^*, k^{**} of $\varkappa$, called splitting and inertial field respectively:

$$\begin{array}{ccccccc} k & \subset & k^* & \subset & k^{**} & \subset & \varkappa \\ & (g) & & (f) & & (e) & \end{array}$$

This scheme of the tower indicates the relative degrees. k^{**}/k^* is a Galois field.

If $\mathfrak{H}$ is the Galois group of $\varkappa$ relative to any subfield k^+ (over k), then it follows at once from the very definition that the splitting and inertial groups $\mathfrak{T}^+$ and

$\mathfrak{T}^+$ of $\mathfrak{P}$ in $\varkappa/k^+$ are the intersections of the full splitting and inertial groups $\mathfrak{Z}$ and $\mathfrak{T}$ with $\mathfrak{h}$. We apply this remark first to k^* as ground field, and in a readily understandable notation then find

$$\mathfrak{Z}^* = \mathfrak{Z}, \quad \mathfrak{T}^* = \mathfrak{T}, \quad \text{thus} \quad e^*f^* = ef, \quad e^* = e.$$

Hence $f^* = f$, and since the degree n^* of $\varkappa/k^*$ equals ef, $g = 1$. Therefore

$$\mathfrak{p}^* = \mathfrak{P}^e \tag{10.3}$$

is a prime ideal in k^*. $\mathfrak{p}$ splits off its first factor

$$\mathfrak{p}_1^* = \mathfrak{P}_1^e$$

when one passes from k to k^*; this explains the names splitting field and splitting group. Since the relative degree f^* of $\mathfrak{P}$ with respect to k^* is the same as its absolute degree f, $\mathfrak{p}^*$ is a prime ideal in k^* of degree 1.

In a second step we take k^{**} as ground field and then find

$$\mathfrak{Z}^{**} = \mathfrak{T}, \quad \mathfrak{T}^{**} = \mathfrak{T}; \quad e^{**}f^{**} = e, \quad e^{**} = e,$$

and hence

$$f^{**} = 1, \quad g^{**} = 1.$$

This proves that (10.3) stays prime in k^{**}, but as the relative degree of $\mathfrak{P}$ in $\varkappa/k^{**}$ is now 1, (10.3) as a prime ideal in k^{**} is of degree f. The effect of the transition from k^* to the higher field is simply the increase of the degree of the prime ideal $\mathfrak{p}^*$ from 1 to f. We summarize:

Theorem III 10, B. *In the splitting field of $\mathfrak{P}$ the prime ideal $\mathfrak{p}$ in k splits off the prime factor $\mathfrak{p}^* = \mathfrak{P}^e$ of degree 1; in passing from the splitting to the inertial field, $\mathfrak{p}^*$ stays prime but its degree increases to f, in passing from the inertial to the full field $\varkappa$, $\mathfrak{p}^*$ breaks up into e equal prime factors $\mathfrak{P}$ of the same degree f.*

As always, let π be a fixed prime number to $\mathfrak{P}$. Hilbert introduced as j^{th} ramification group $\mathfrak{V}_j$ that subgroup of $\mathfrak{T}$ whose elements v satisfy the congruence

$$\pi^v \equiv \pi \quad (\mathfrak{P}^{j+1}).$$

$\mathfrak{T}$ itself is $\mathfrak{V}_0$, and $\mathfrak{V}_j$ is subgroup of $\mathfrak{V}_{j-1}$. The corresponding subfields of $\varkappa$ over k^{**} are the ramification fields

$$k^{(0)} = k^{**},\ k^{(1)},\ k^{(2)},\ \ldots .$$

The sequence breaks off because for a sufficiently high j the group $\mathfrak{V}_j$ consists of the unit element only. The representation (10.2) at once shows that the element v of $\mathfrak{G}$ lies in $\mathfrak{V}_j$ if and only if

$$\alpha^v \equiv \alpha \quad (\mathfrak{P}^{j+1})$$

for all integers α in $\varkappa$. By the same argument as for $\mathfrak{V}_0$ one finds that $\mathfrak{V}_0, \mathfrak{V}_1, \ldots$ are all invariant subgroups of $\mathfrak{G}$, or $k^{(0)}, k^{(1)}, k^{(2)}, \ldots$ are Galois fields over the splitting field k^*. A fortiori, each of the extensions

$$k^{(0)}/k^*,\quad k^{(1)}/k^{(0)},\quad k^{(2)}/k^{(1)},\ \ldots$$

is Galois. I maintain that all, with the possible exception of the first one, are Abelian, i.e., that their groups $\mathfrak{V}_1, \mathfrak{V}_2, \ldots$ are commutative. (In case of numerical fields, the first one, $k^{(0)}/k^*$ also is Abelian and even cyclic.)

Indeed, for any t in $\mathfrak{T}$

$$\frac{\pi^t}{\pi} \equiv \rho_t \ (\mathfrak{P})$$

with a certain unit ρ_t in $\varkappa_{\mathfrak{P}}$. Two t are in the same coset mod $\mathfrak{V}_1$ if and only if their ρ_t coincide mod $\mathfrak{P}$. For t,t' in $\mathfrak{T}$ one finds

$$\frac{\pi^{tt'}}{\pi^{t'}} \equiv \rho_t^{t'} \equiv \rho_t \ (\mathfrak{P})$$

and

$$\frac{\pi^{tt'}}{\pi} \equiv \rho_t \rho_{t'} \ (\mathfrak{P}) \quad \text{or} \quad \rho_{tt'} \equiv \rho_t \rho_{t'} \ (\mathfrak{P})$$

Therefore $\mathfrak{V}_0/\mathfrak{V}_1$ is isomorphic to a subgroup of the multiplicative group of the residues prime to $\mathfrak{P}$.

All substitutions t in $\mathfrak{V}_1$ satisfy

$$\frac{\pi^t}{\pi} \equiv 1 \ (\mathfrak{P}), \quad \left(\frac{\pi^t}{\pi}\right)^h \equiv 1 \ (\mathfrak{P})$$

or for any exponent h = 1, 2,...

(10.4) $$(\pi^t)^h \equiv \pi^h \ (\mathfrak{P}^{h+1}).$$

Consider now $\mathfrak{V}_j/\mathfrak{V}_{j+1}$ for $j \geqq 1$ and let v be any element of $\mathfrak{V}_j$. We write

(10.5) $$\frac{\pi^v}{\pi} \equiv 1 + \sigma_v \pi^j \ (\mathfrak{P}^{j+1})$$

where σ_v is an element of $\varkappa_{\mathfrak{P}}$. Two v are in the same coset mod $\mathfrak{V}_{j+1}$ if and only if their σ_v coincide mod $\mathfrak{P}$. For u,v in $\mathfrak{V}_j$ one derives from (10.5) by the substitution u:

$$\frac{\pi^{vu}}{\pi^u} \equiv 1 + \sigma_v^u (\pi^u)^j \ (\mathfrak{P}^{j+1}),$$

and, because of $\sigma_v^u \equiv \sigma_v \ (\mathfrak{P})$ and (10.4), this is

$$\equiv 1 + \sigma_v \pi^j \ (\mathfrak{P}^{j+1}).$$

In multiplying by π^u/π one gets

$$\pi^{vu} \equiv 1 + (\sigma_u + \sigma_v)\pi^j \ (\mathfrak{P}^{j+1}) \quad \text{or} \quad \sigma_{vu} \equiv \sigma_u + \sigma_v \ (\mathfrak{P}).$$

Hence $\mathfrak{V}_j/\mathfrak{V}_{j+1}$ $(j \geqq 1)$ is isomorphic to a subgroup of the <u>additive</u> group of <u>all</u> residues mod $\mathfrak{P}$.

For numerical fields the degree of $\mathfrak{V}_0/\mathfrak{V}_1$ must be a divisor of $(N\mathfrak{y}) - 1$, and the degrees of $\mathfrak{V}_1/\mathfrak{V}_2, \mathfrak{V}_2/\mathfrak{V}_3, \ldots$ are divisors of $N\mathfrak{y}$ and thus powers of the prime characteristic p of $k_{\mathfrak{y}}$.

The exact power $\mathfrak{P}^{\varepsilon-1}$ of $\mathfrak{P}$ which goes into the ramification ideal of $\varkappa/k$ is by definition the order of the product

$$\prod_t (\pi - t\pi) \qquad (t \text{ in } \mathfrak{T} \text{ and } \neq 1).$$

This shows anew that $\varepsilon \geqq e$, but enables one to determine ε completely in terms of the degrees $w_0 = e, w_1, \ldots$ of the ramification groups $\mathfrak{V}_0, \mathfrak{V}_1, \ldots$. Indeed $\pi - t\pi$ is of the order j + 1 if t lies in $\mathfrak{V}_j$ but not in $\mathfrak{V}_{j+1}$. Hence

$$\varepsilon - 1 = (w_0 - 1) + (w_1 - 1) + (w_2 - 1) + \ldots ,$$

a relation conveniently to be compared with

$$e = \frac{w_0}{w_1} \cdot \frac{w_1}{w_2} \cdots .$$

In case of a numerical field the first factor is prime to p, the others are powers of p; hence $\varepsilon > e$ if and only if $e : p$. This confirms and sharpens our former results about the ramification ideal.

Theorem III 10, C. *All ramification groups $\mathfrak{W}_0 = \mathfrak{T}, \mathfrak{W}_1, \ldots$ are invariant subgroups of $\mathfrak{Y}$. The factor group $\mathfrak{W}_0/\mathfrak{W}_1$ is isomorphic with a subgroup of the multiplicative group of the residues prime to $\mathfrak{P}$, whereas $\mathfrak{W}_1/\mathfrak{W}_2, \mathfrak{W}_2/\mathfrak{W}_3, \ldots$ are isomorphic with subgroups of the additive group of all residues mod $\mathfrak{P}$. With $w_0, w_1, \ldots$ being the degrees of the successive ramification groups, one has*

$$\varepsilon - 1 = (w_0 - 1) + (w_1 - 1) + \cdots .$$

We return once more to the case of a numerical field. If $\mathfrak{P}$ is not a divisor of the ramification ideal of $\varkappa/k$, then the substitution s of the splitting group of $\mathfrak{P}$ for which all integers α satisfy the congruence

$$(10.6) \qquad \alpha^s \equiv \alpha^P \ (\mathfrak{P}) \qquad [P = N\mathfrak{p}]$$

is uniquely determined and not merely modulo the inertial group. We denote this substitution of the Galois group, known as the Frobenius substitution of $\mathfrak{P}$, by

$$s = \left(\frac{\varkappa}{\mathfrak{P}}\right).$$

Its order f equals the relative degree f of $\mathfrak{P}$. Let u be any element of the Galois group. From (10.6) one infers

$$\alpha^{su} \equiv (\alpha^u)^P \ (\mathfrak{P}^u) \quad \text{or} \quad \beta^{u^{-1}su} \equiv \beta^P \ (\mathfrak{P}^u)$$

with $\beta = \alpha^u$, which shows that $u^{-1}su$ is the Frobenius substitution of $\mathfrak{P}^u$:

$$\left(\frac{\varkappa}{\mathfrak{P}^u}\right) = u^{-1} \cdot \left(\frac{\varkappa}{\mathfrak{P}}\right) \cdot u.$$

In particular, if $\varkappa$ is an <u>Abelian field</u> over k, whose Galois group is commutative, then the Frobenius substitution s is the same for all the conjugates $\mathfrak{P}^u$ and depends only on the prime ideal $\mathfrak{p}$ in k whose factors in $\varkappa$ they are. We denote it therefore by

(10.7) $$\left(\frac{\varkappa}{\mathfrak{p}}\right)$$

Since $\mathfrak{p} = \mathfrak{P}_1 \ldots \mathfrak{P}_g$ splits into distinct prime ideals $\mathfrak{P}_1, \ldots, \mathfrak{P}_g$, (10.6) results in

(10.8) $$\alpha^s \equiv \alpha^P \ (\mathfrak{p}).$$

The Artin symbol (10.7) associates a substitution of the Galois group of $\varkappa/k$ with the prime ideal $\mathfrak{p}$ of k, provided $\mathfrak{p}$ does not divide the discriminant $\mathfrak{d}$ of $\varkappa/k$. For any integral or fractional ideal

$$\mathfrak{a} = \mathfrak{p}_1^{h_1} \mathfrak{p}_2^{h_2} \cdots$$

we set

$$\left(\frac{\varkappa}{\mathfrak{a}}\right) = \left(\frac{\varkappa}{\mathfrak{p}_1}\right)^{h_1}\left(\frac{\varkappa}{\mathfrak{p}_2}\right)^{h_2} \cdots$$

so that

$$\left(\frac{\varkappa}{\mathfrak{a}\mathfrak{b}}\right) = \left(\frac{\varkappa}{\mathfrak{a}}\right) \cdot \left(\frac{\varkappa}{\mathfrak{b}}\right).$$

The fractional ideals $\mathfrak{a}$ whose numerator and denominator are prime to $\mathfrak{d}$ form a group; those among them which satisfy the equation $\left(\frac{\varkappa}{\mathfrak{a}}\right) = 1$ form a subgroup of finite index. The factor group is the group of classes of ideals if we reckon in the same class two ideals $\mathfrak{a}$, $\mathfrak{b}$ (prime to $\mathfrak{d}$) satisfying the equation $\left(\frac{\varkappa}{\mathfrak{a}}\right) = \left(\frac{\varkappa}{\mathfrak{b}}\right)$. The relationship

(10.9) $$\mathfrak{a} \to \left(\frac{\varkappa}{\mathfrak{a}}\right)$$

establishes an isomorphic mapping of the class group of ideals into the Galois group of $\varkappa/k$. It has been found that the Artin symbol lies at the root of the higher reciprocity laws, and the correspondence (10.9) is the starting point of the theory of class fields, the deepest and most advanced part of the arithmetic of algebraic numbers. Cf. the last two sections of Ch. IV.

Here we consider as an example the l-cyclotomic field over the rational field $\mathfrak{g}$, l being a prime. Let $\mathfrak{p}$ be a prime factor in $\mathfrak{g}(\zeta)$ of the rational prime number $p \neq l$. The substitution s of the Galois group for which $\alpha^s \equiv \alpha^p$ $(\mathfrak{p})$ is defined by $\zeta \to \zeta^p$. Hence if a is an arbitrary rational integer $\not\equiv 0(l)$, one finds

$$\left(\frac{\mathfrak{g}(\zeta)}{a}\right) = (\zeta \to \zeta^a).$$

The distribution of the numbers a which are prime to l into classes coincides with their distribution into the l - 1 different congruence classes, and the cyclotomic field $\mathfrak{g}(\zeta)$ is a class field for this grouping.

11. Cyclotomic Field and Quadratic Law of Reciprocity

It is the common curse of all general and abstract theories that they have to be far advanced before yielding useful results in concrete problems. The next two sections are proof that we have reached that level.

With the odd prime number l we form the l-cyclotomic field $\mathfrak{g}(\zeta)$ of degree l - 1 over $\mathfrak{g}$. The residues mod l form the simplest strictly finite field; we have seen in Chap. I, §4, that its elements after the exclusion of zero form a cyclic group under multiplication, i.e., there exists a primitive residue r such that all residues $\not\equiv 0(l)$ are contained in the series

$$1, r, \ldots, r^{l-2}.$$

The Galois group of $\mathfrak{g}(\zeta)$ consists of the substitutions

$$\zeta \to \zeta^g \qquad (g = 1, \ldots, l - 1)$$

and is isomorphic to the multiplicative group of the residues g. Hence it is also cyclic, with s: $\zeta \to \zeta^r$ as a generator.

$1, s^2, s^4, \ldots, s^{l-3}$ form a subgroup of index 2; therefore $\mathfrak{g}(\zeta)$ contains a subfield k of degree 2. Its discriminant d, as we saw in §1, contains (1) no odd prime to a power higher than the first, (2) the prime 2 either not at all or in the second power, and under the first alternative d will be $\equiv 1$ (4). As the discriminant of $\mathfrak{g}(\zeta)$ is a power of l and the discriminant of the subfield k is a divisor thereof, we see that d must be $\pm l$ with such sign that

$\pm l \equiv 1$ (4) and the quadratic field k is $\mathfrak{g}(\sqrt{\pm l})$. One may write more explicitly

$$\pm l = (-1)^{\frac{l-1}{2}} \cdot l.$$

From the factorization of a prime number $p \neq l$ in $\mathfrak{g}(\zeta)$ we should be able to derive its factorization in the quadratic subfield. We have learned that

$$(p) = \mathfrak{p}_1 \ldots \mathfrak{p}_g \quad \text{in } \mathfrak{g}(\zeta)$$

where $\mathfrak{p}_1, \ldots, \mathfrak{p}_g$ are distinct prime ideals of degree f, $gf = l - 1$, and f is the least exponent satisfying the congruence

(11.1) $$p^f \equiv 1 \pmod{l}.$$

$$\mathfrak{p}_j = \mathfrak{p}^{s^{j-1}} \; (j = 1, \ldots, g); \quad \mathfrak{p}^{s^g} = \mathfrak{p}.$$

$1, s^g, \ldots, s^{(f-1)g}$ constitute the splitting group. If p decomposes into two distinct prime ideals $\mathfrak{q}$, $\mathfrak{q}' = \mathfrak{q}^s$ in k, it is evident that g must be even, since $\mathfrak{q}$ and $\mathfrak{q}'$ break into the same number of prime factors in $\mathfrak{g}(\zeta)$:

$$\mathfrak{q} = \mathfrak{p}^{1+s^2+\ldots},$$
$$\mathfrak{q} = \mathfrak{p}^{s+s^3+\ldots}.$$

[Hilbert introduced the habit of writing substitutions as exponents in order to make it possible to write a product

$$\gamma^{c_0}(s\gamma)^{c_1}(s^2\gamma)^{c_2} \ldots$$

with integral exponents c as $\gamma^{F(s)}$ with the symbolic polynomial

$$F(s) = c_0 + c_1 s + c_2 s^2 + \ldots .]$$

Vice versa, if g is even, the splitting group of $\mathfrak{p}$ in $\mathfrak{g}(\zeta)/k$ is the same as in $\mathfrak{g}(\zeta)/\mathfrak{g}$, that is to say, f does not change and therefore g is halved so that $\mathfrak{q}$ contains only $\frac{1}{2}g$ prime ideals in $\mathfrak{g}(\zeta)$ and one finds $(p) = \mathfrak{q}\mathfrak{q}'$. Thus

$$(p) = \mathfrak{q}\mathfrak{q}' \quad \text{or} \quad (p) = \mathfrak{q} \quad \text{in k}$$

according as g is even or odd, i.e., according as $\frac{1}{2}(l - 1)$ is or is not a multiple of f. Hence our criterion boils down to the alternative

$$(11.2) \qquad p^{\frac{1}{2}(l-1)} \equiv +1 \quad \text{or} \quad -1 \quad (\text{mod } l).$$

Here enters a well-known elementary consideration about quadratic residues. Fermat's theorem for an integer $a \not\equiv 0 \ (l)$ states that

$$(a^{\frac{l-1}{2}} - 1)(a^{\frac{l-1}{2}} + 1) \equiv 0 \quad (l).$$

The quadratic residues mod l obviously satisfy the congruence

$$a^{\frac{l-1}{2}} \equiv 1 \quad (l),$$

Since it cannot have more than $\frac{l - 1}{2}$ incongruent roots, and this is exactly the number of quadratic residues, all the non-residues satisfy the other congruence

$$a^{\frac{l-1}{2}} \equiv -1 \quad (l).$$

Consequently

$$(11.3) \qquad a^{\frac{l-1}{2}} \equiv \left(\frac{a}{l}\right) \qquad (\text{mod } l).$$

This argument yields the so-called first supplement of the reciprocity law:

$$(11.4) \quad \left(\frac{-1}{l}\right) = \pm 1 \quad \text{according as} \quad l \equiv \pm 1 \quad (\text{mod } 4).$$

It changes the alternative (11.2) above attained into the statement that p splits or does not split in k according as

$$\left(\frac{p}{l}\right) = +1 \quad \text{or} \quad -1.$$

On the other hand we are in possession of a criterion for the splitting of a prime number p in a quadratic field $\mathcal{g}(\sqrt{a})$. By applying this to our present field $\mathcal{g}(\sqrt{\pm l})$ we obtain

(1) for p = 2:

$$(11.5)\qquad \begin{cases} \left(\frac{2}{l}\right) = 1 & \text{if } l \equiv \pm 1 \mod 8, \\ \left(\frac{2}{l}\right) = -1 & \text{if } l \equiv \pm 5 \mod 8; \end{cases}$$

(2) for an odd prime p:

$$(11.6)\qquad \left(\frac{p}{l}\right) = \left(\frac{\pm l}{p}\right).$$

This is Gauss' famous reciprocity law with its second supplement. In making use of the equation (11.4) for p rather than l, the reciprocity law (11.6) may be put into its usual form:

$$\left(\frac{p}{l}\right) = \left(\frac{l}{p}\right) \text{ if } l \text{ or } p \equiv 1 \pmod 4,$$

$$\left(\frac{p}{l}\right) = -\left(\frac{l}{p}\right) \text{ if both } l \text{ and } p \equiv 3 \pmod 4.$$

There certainly exist more elementary proofs of the reciprocity law, but hardly one that is less artificial and goes as straight to the root of the phenomenon.

We determined the quadratic subfield of $\mathfrak{g}(\zeta)$ by means of the general theorem III 9, C, on discriminants. Of course there is a more elementary way by means of the Gaussian sums. Let a, b range over the quadratic residues and non-residues mod l.

$$\eta = \sum_a \zeta^a$$

is a number of k and

$$\eta' = \sum_b \zeta^b$$

its conjugate. The coefficient 0 of x^{l-1} in the polynomial $x^l - 1$ shows that

$$1 + \eta + \eta' = 0.$$

To compute

$$\eta\eta' = \sum_{a,b} \zeta^{a+b}$$

one has to ascertain how often a given residue g may be represented as a + b. We first observe that this will occur for any residue $g \not\equiv 0\ (l)$ as often as for g = 1. Indeed, each decomposition of 1 into a sum of a residue and a non-residue,

$$1 = a + b,$$

leads to a corresponding decomposition of g,

$$g = a' + b', \tag{11.7}$$

by

$a' = ga,\quad b' = gb$ provided g is quadr. res.,

$b' = ga,\quad a' = gb$ " g " " non-res.

If -1 is a quadratic residue, the equation a + b = 0 is impossible. The total number of pairs (a,b) amounts to $\left(\frac{l-1}{2}\right)^2$. Hence every residue g = 1,..., l - 1 appears exactly $\frac{l-1}{4}$ times, and as a by-product we find again that in this case $l \equiv 1\ (4)$.

If, however, -1 is a quadratic non-residue, then a + b = 0 has $\frac{l-1}{2}$ solutions, namely: a an arbitrary quadratic residue and b = -a. Hence the other residues g = 1,..., l - 1 are represented in the form (11.7) each $\frac{l-3}{4}$ times, and in this case one must have $l \equiv 3\ (4)$. This simple argument involves a new proof of the first supplementary law (11.4)

In the first case we find

$$\eta\eta' = \frac{l-1}{4} \cdot \sum_{g \not\equiv 0(l)} \zeta^g = \frac{1-l}{4},$$

in the second case

$$\eta\eta' = \frac{l-1}{2} + \frac{l-3}{4} \cdot \sum_{g \not\equiv 0(l)} \zeta^g = \frac{1+l}{4}.$$

The resulting quadratic equations are

$$\eta^2 + \eta + \frac{1 - l}{4} = 0, \qquad \eta^2 + \eta + \frac{1 + l}{4} = 0$$

and their solutions

$$\eta = \frac{-1 \pm \sqrt{l}}{2} \quad \text{and} \quad \eta = \frac{-1 \pm \sqrt{-l}}{2} \tag{11.8}$$

respectively.

12. General Cyclotomic Fields

If k is a numerical Dedekind field, and $\mathfrak{p}$ a prime ideal, we denote by $P = N\mathfrak{p}$ the number of elements in $k_{\mathfrak{p}}$, i.e., of residues mod $\mathfrak{p}$. P is a power of the prime characteristic p of $k_{\mathfrak{p}}$. Every integer a in k satisfies the congruence

$$a^P \equiv a \quad (\text{mod } \mathfrak{p}).$$

If $\varkappa$ is a finite field over k, $\mathfrak{P}$ a prime divisor of $\mathfrak{p}$ in $\varkappa$, and $\varkappa_{\mathfrak{P}}/k_{\mathfrak{p}}$ of (Kummer) degree f, one has

$$N\mathfrak{P} = (N\mathfrak{p})^f. \tag{12.1}$$

The fact that the Krnoecker degree of $\mathfrak{P}$ equals its Kummer degree is expressed by the equation for the relative norm of $\mathfrak{P}$:

$$\text{Nm}\, \mathfrak{P} = \mathfrak{p}^f. \tag{12.2}$$

In particular, for any finite field $\varkappa$ over $\mathfrak{g}$ and a prime ideal $\mathfrak{p}$ of degree f dividing the rational prime number p:

$$N\mathfrak{p} = p^f, \qquad \text{Nm}\, \mathfrak{p} = (p)^f. \tag{12.3}$$

After this preliminary remark we turn to our subject proper, the cyclotomic fields of arbitrary degree. The equation

$$x^m = 1$$

has in Ω the m roots

$$1, Z, \ldots, Z^{m-1} \quad \text{with} \quad Z = e^{2\pi i/m}$$

which form the vertices of a regular m-gon on the unit circle. They (or the rotations which carry the regular m-gon into itself) form a multiplicative group isomorphic to the additive group of residues mod m. If one winds a straight line on which the integers g are marked upon a circle of circumference m, two marks g,g' will automatically coincide if $g \equiv g'$ (m), and one thus achieves the transition from the ring of integers to the strictly finite ring of "residues modulo m." The process is described analytically by the correspondence

$$g \rightarrow Z^g.$$

It is not surprising therefore that the m-cyclotomic field $\mathcal{g}(Z)$ is an important tool for the investigation of common integers mod m.

For a purely algebraic construction of the cyclotomic field one conveniently starts with a prime power $m = l^h$. As ground level will serve any finite field k over $\mathcal{g}$. We first develop a few auxiliary formulas. Setting for the moment

$$x_0 = x, \quad x_1 = x^l, \quad x_2 = x^{l^2}, \ldots$$

one has

$$(12.4) \quad x_h - 1 = (x_{h-1} - 1)(x_{h-1}^{l-1} + \ldots + x_{h-1} + 1)$$

$$= (x_{h-1} - 1) \cdot f(x).$$

The polynomial f(x) is of degree

$$\varphi(l^h) = l^{h-1}(l - 1) = L.$$

An irreducible factor $f^*(x)$ of f(x) in k serves to define the cyclotomic extension K = k(Z) of k:

$$f^*(Z) = 0, \quad \text{hence} \quad f(Z) = 0, \quad Z^{l^h} = 1.$$

The least exponent d for which $Z^d = 1$ must be a divisor of l^h. Hence either $d = l^h$ or

$$Z^{l^{h-1}} = 1.$$

The second alternative would imply $f(Z) = 1$ and is thus excluded. Consequently Z is a primitive l^hth root of unity, i.e., the l^h roots

$$Z^g \quad (g = 0, 1, \ldots, l^h - 1)$$

are distinct and we have in $k(Z)$:

(12.5) $x^{l^h} - 1 = \prod_g (x - Z^g)$ (g all residues mod l^h),

(12.6) $f(x) = \prod_g (x - Z^g)$ (g all such residues prime to l).

Z is a unit in K. We put

$$\Lambda = 1 - Z, \quad \mathcal{L} = (\Lambda).$$

For any primitive

$$Z' = Z^g, \qquad g \not\equiv 0 \ (l),$$

the quotient

$$\frac{1 - Z'}{1 - Z} = 1 + Z + \ldots + Z^{g-1}$$

is integral, but because of the primitivity of Z',

$$Z = Z'^{g'}, \qquad gg' \equiv 1 \ (l^h),$$

the inverse quotient also is integral. Consequently all factors of the right member of the equation arising from (12.6) by the substitution $x = 1$,

(12.7) $$l = \prod_g (1 - Z^g) \qquad \{g \not\equiv 0 \ (l)\}$$

are $\sim\Lambda$ and we obtain

$$(l) = \mathcal{L}^L.$$

If we express the first factor at the right member of (12.4) by the same formula for $h - 1$ instead of h, continue downwards and finally substitute $x = 1$, then we find the following equation similar to (12.7):

$$l^h = \prod_g (1 - Z^g) \quad \text{(g all residues except 0).}$$

An immediate consequence is the

Lemma III 12, A. *If $\mathfrak{p}$ is a prime ideal in K not dividing l, a congruence*

$$Z^a \equiv Z^b \ (\mathfrak{p})$$

necessarily entails the equation

$$Z^a = Z^b \quad or \quad a \equiv b \ (l^h).$$

By deriving (12.4) one finds for the differential $\Delta = \dot{f}(Z)$ the relation

$$l^h \cdot Z^{l^h - 1} = (\zeta - 1)\Delta$$

where $\zeta = Z^{l^{h-1}}$ is a primitive l^{th} root of unity; hence

$$\Delta = \frac{l^h}{Z(\zeta - 1)} . \tag{12.8}$$

We now introduce an essential assumption about k and then prove a string of propositions about k(Z).

Hypothesis H: l splits in k into distinct prime ideals,

$$(l) = \mathfrak{l}_1 \mathfrak{l}_2 \ldots \mathfrak{l}_u.$$

Theorem III 12, B. *Under the hypothesis H, f(x) is irreducible in k and*

$$1, Z, \ldots, Z^{L-1} \tag{12.9}$$

constitute an integral basis for the field k(Z)/k of degree L.

The proof is suggested by our procedure in §4 for the l-cyclotomic field. Put

$$\mathfrak{L}_1 = (\mathfrak{L}, \mathfrak{l}_1).$$

Then

$$\mathfrak{L}_1^L = (\mathfrak{L}^L, \mathfrak{l}_1^L) = (l, \mathfrak{l}_1^L) = (\mathfrak{l}_1 \mathfrak{l}_2 \cdots, \mathfrak{l}_1^L) = \mathfrak{l}_1.$$

This equation does not allow the relative degree of K/k to sink below L. Hence the degree is exactly L and $\mathcal{L}_1$ a prime ideal in K of relative degree 1.

$$(l) = \mathcal{L}^L = \ell_1 \dots \ell_u = (\mathcal{L}_1 \dots \mathcal{L}_u)^L$$

implies

$$\mathcal{L} = \mathcal{L}_1 \dots \mathcal{L}_u.$$

K is a Galois field over k, and

$$s : Z \to Z^g \quad (g \text{ prime to } l)$$

are its L automorphisms.

Our next concern is the discriminant of the basis (12.9) which (but for the sign) equals the norm of $\Delta = \dot{f}(Z)$. As one readily sees from (12.8) the norm has the value:

$$(12.10) \qquad l \text{ to the power } l^{h-1}(hl - h - 1) = L(h - \frac{1}{l - 1}).$$

The important point is that the discriminant is a power of l.

By an argument employed before in §4, the second part of our theorem is therefore reduced to the following

Lemma III 12, C. *If*

$$\alpha_0 + \alpha_1\Lambda + \dots + \alpha_{L-1}\Lambda^{L-1}$$

has integral coefficients α_i *in k and is itself divisible by l, then each coefficient is divisible by l.*

The assumption implies $\alpha_0 : \mathcal{L}$, hence $\alpha_0 : \mathcal{L}_1$, and since α_0 is in k, $\alpha_0 : \ell_1$. [Indeed $(\alpha_0, \ell_1) = 1$ is precluded by the common divisor $\mathcal{L}_1$ of α_0 and ℓ_1 in K.] Similarly $\alpha_0 : \ell_2, \dots$, therefore $\alpha_0 : l$. After subtraction of α_0 one finds

$$\alpha_1 + \alpha_2\Lambda + \dots + \alpha_{L-1}\Lambda^{L-2}$$

to be still divisible by $\mathcal{L}^{L-1}$, and the above argument may be repeated until the series of coefficients is exhausted.

Theorem III 12, D. *Under the Hypothesis H, the prime ideal* ℓ_1 *going into l splits in K into L equal prime factors* $\mathfrak{L}_1$ *of relative degree 1; the same for* $\mathfrak{l}_2, \dots, \mathfrak{l}_u$.

A prime ideal $\mathfrak{p}$ *of k which does not divide l, splits in K into g distinct prime ideals of the same relative degree f,*

$$\mathfrak{p} = \mathfrak{P}_1 \cdots \mathfrak{P}_g.$$

f is the least exponent for which

$$(N\mathfrak{p})^f \equiv 1 \ (l^h)$$

and $fg = L$.

The first part has already been proved; the second could be settled by the same method as employed in §4 for the l-cyclotomic field. We prefer to use the Galois theory of §10. Let $\mathfrak{p}$ be a prime ideal of k not dividing l, and $\mathfrak{P}$ a prime divisor of $\mathfrak{p}$ in K. If s is a substitution of the inertial group of $\mathfrak{P}$, we must have

$$Z^s = Z^g \equiv Z \ (\mathfrak{P}).$$

Hence by Lemma III 12, A, $Z^g = Z$ or $s = 1$. The inertial group is thus of degree 1, $e = 1$, and the splitting group coincides with the Galois group of $K_{\mathfrak{P}}/k_{\mathfrak{p}}$ which consists of the automorphisms

$$A \rightarrow A, \quad A \rightarrow A^P, \dots, A \rightarrow A^{P^{f-1}} \qquad (P = N\mathfrak{p}).$$

f is the least exponent such that

$$A^{P^f} \equiv A \quad (\text{mod } \mathfrak{P}) \tag{12.11}$$

for every integer A in K. At this point we avail ourselves of the integral basis (12.9) and by means of the representation

$$A = \alpha_0 + \alpha_1 Z + \dots + \alpha_{L-1} Z^{L-1}$$

with integral components α in k we realize that the congruence (12.11) prevails for all integers if and only if it holds for $A = Z$. Hence f is the least exponent for which

$$Z^{P^f} \equiv Z \ (\mathfrak{P})$$

or, according to Lemma III 12, A,

$$p^f \equiv 1\ (l^h).$$

(The absence of multiple prime divisors of $\mathfrak{p}$ in K could also have been derived from Dedekind's theorem about prime divisors of the discriminant.)

The two theorems III 12, B and D, enable us to construct the m-cyclotomic field by consecutive adjunctions. We factorize m into powers of distinct primes,

$$m = l_1^{h_1} \ldots l_t^{h_t},$$

and build the t-story tower

$$(12.12) \quad k_0 = \mathcal{R},\ k_1 = k_0(Z_1), \ldots,\ k_j = k_{j-1}(Z_j), \ldots,\ k_t = K$$

where the extension $k_{j-1} \rightarrow k_j$ takes place according to the pattern above described as $k \rightarrow k(Z)$ with $l_j^{h_j}$ for l^h. The second part of Theorem III 12, D gives the key to the induction. Its validity for k_{j-1} lays the foundation for the jth story by furnishing the Hypothesis H for the adjunction of Z. More explicitly: the statement that any rational prime number $p \neq l_1, \ldots, l_j$ splits in k_j into distinct prime ideals is proved by induction with respect to j. Assuming its truth for j - 1, one knows that l_j splits in k_{j-1} into distinct prime factors. Hence Hypothesis H is fulfilled for the adjunction of Z ; and if p is not only $\neq l_1, \ldots, l_{j-1}$ but also $\neq l_j$, then it splits in k_{j-1} into distinct prime ideals

$$(p) = \mathfrak{p}_1 \cdots \mathfrak{p}_g$$

and, according to Theorem III 12, D, each of the factors $\mathfrak{p}_1, \ldots, \mathfrak{p}_g$ in its turn splits in k_j into distinct prime factors, which proves our statement for j. [Observe that two distinct prime ideals $\mathfrak{p}, \mathfrak{q}$ in a field k can have no common prime divisor in a field K over k since $(\mathfrak{p}, \mathfrak{q}) = 1$.] We are sure that the jth extension in the sequence (12.12) takes place by means of the polynomial

$$\sum_{i=0}^{l-1} x^{il^{h-1}}$$

irreducible in k_{j-1}. The most important result of this

induction is the fact that K is of degree

$$L_1 \ldots L_t = \varphi(m)$$

(= number of residues prime to m).

We write

$$m = l_1^{h_1} \cdot m_1 = \ldots = l_t^{h_t} \cdot m_t.$$

The number

$$Z = Z_1 \ldots Z_t$$

in K is an mth root of unity,

$$Z^m = 1.$$

I maintain that it is a primitive root. Indeed

$$Z^d = 1$$

implies

$$Z^{m_1 d} = Z_1^{m_1 d} = 1,$$

consequently

$$m_1 d \equiv 0 \ (l_1^{h_1}), \quad d \equiv 0 \ (l_1^{h_1}).$$

For the same reason d is divisible by $l_2^{h_2}, \ldots, l_t^{h_t}$, hence by m. Vice versa $Z_1, \ldots, Z_t$ are expressible as powers of Z:

$$Z_1 = Z^{m_1 m_1'} \quad \text{if} \quad m_1 m_1' \equiv 1 \ (l_1^{h_1}).$$

Hence K = $\mathfrak{g}(Z)$. In k one has

$$x^m - 1 = \prod_g (x - Z^g) \quad (g = 0, 1, \ldots, m - 1). \tag{12.13}$$

The field equation f(x) satisfied by Z is irreducible of degree $\varphi(m)$ and has integral rational coefficients. Being a divisor of $x^m - 1$, f(x) must be a product of some of the linear factors in (12.13):

$$f(x) = \Pi(x - Z^g)$$

(g ranging in a subset $\mathfrak{g}$ of residues mod m).

Hence K/k is a Galois field and

$$s : Z \to Z^g \quad (g \text{ in } \mathcal{G})$$

are its automorphisms. Each has an inverse and g in $\mathcal{G}$ is therefore prime to m. Because of the degree $\varphi(m)$, $\mathcal{G}$ must contain all relatively prime residues, and we find

$$f(x) = \prod_g (x - Z^g)$$

the product extending over the $\varphi(m)$ primitive roots Z^g.

By climbing the tower from floor to floor we obtain an integral basis for K of the telescopic structure

$$Z_1^{a_1} \dots Z_t^{a_t}.$$

But all these products are powers of Z so that every integer A in K is expressible as a polynomial in Z with integral coefficients in $\mathfrak{g}$. Using f(x), one can reduce its degree to $\varphi(m) - 1$, and thus one finds

$$1, Z, \dots, Z^{\varphi(m)-1}$$

to be an integral basis for K.

The whole argument goes through even if the ground level $\mathfrak{g}$ is replaced by a finite field k over $\mathfrak{g}$ in which each of the primes $l_1, \dots, l_t$ splits into distinct prime ideals (Hypothesis H_t):

$$l_1 = \mathfrak{l}_1 \mathfrak{l}_1' \dots ,$$
$$\dots\dots\dots$$
$$l_t = \mathfrak{l}_t \mathfrak{l}_t' \dots .$$

[According to Dedekind's theorem on prime divisors of the discriminant, Hypothesis H_t is equivalent to requiring the discriminant of k to be prime to m.] By an easy integration of the consecutive steps, we obtain the following conclusive result:

Theorem III 12, E. *If the algebraic number field k satisfies the Hypothesis H_t, then the m-cyclotomic extension $k(Z) = K$ is a Galois field of relative degree $\varphi(m)$ whose Galois group is isomorphic with the multiplicative group of the residues prime to m, and*

$$1, Z, \dots, Z^{\varphi(m)-1}$$

is an integral basis for K/k.

A prime ideal $\mathfrak{y}$ *of k which is contained in no one of the prime numbers* $l_1, \ldots, l_t$ *splits in K into* g *distinct prime ideals of the same degree f relative to k:*

$$\mathfrak{y} = \mathfrak{P}_1 \cdots \mathfrak{P}_g.$$

f is the least exponent such that

$$(N\mathfrak{y})^f \equiv 1 \ (m)$$

and $fg = \varphi(m)$. *The prime ideals* $\mathfrak{l}_j, \mathfrak{l}_j', \ldots$ *of* l_j *split according to the scheme*

$$\mathfrak{l}_j = (\mathfrak{L}_{j1} \cdots \mathfrak{L}_{j,g_j})^{L_j}, \quad L_j = l_j^{h_j - 1}(l_j - 1),$$

where $\mathfrak{L}_{j1}, \ldots, \mathfrak{L}_{jg_j}$ *are distinct prime ideals of the same degree* f_j. *This degree is the least exponent such that*

$$(N\mathfrak{l}_j)^{f_j} \equiv 1 \ (m_j)$$

and $f_j g_j = \varphi(m_j)$.

(If we adjoin $Z_1, \ldots, Z_t$ in this order, the decomposition of $\mathfrak{l}_t, \mathfrak{l}_t', \ldots$ is readily found to be in agreement with the above statement. However, within K, and starting from the ground level k, we may carry out the adjunction in any order. What holds good for l_t must therefore be true for $l_1, \ldots, l_{t-1}$.)

It is easy to compute the ramification ideal $\mathfrak{d}$; the result is

$$\mathfrak{d} = \left(\frac{m}{(1 - \zeta_1) \cdots (1 - \zeta_t)}\right)$$

where $\zeta_1, \ldots, \zeta_t$ are primitive $l_1, \ldots, l_t^{th}$ roots of unity.

In the multiplicative group of the mth roots of unity,

$$Z' = 1, Z, \ldots, Z^{m-1},$$

each element has its order d which is a divisor of m, and the elements of order d are

(12.14) $$Z_{(d)}^{i} \quad \text{(i residue prime to d)}$$

with $Z_{(d)} = Z^{m/d}$. If we introduce the polynomial $f_d(x)$ of degree $\varphi(d)$ whose zeros are the roots (12.14) of exact order d, then we must have

(12.15) $$x^m - 1 = \prod_{d|m} f_d(x),$$

the symbol $d|m$ indicating that the product extends over all divisors d of m. For the same reason

(12.16) $$x^d - 1 = \prod_{\delta|d} f_\delta(x) \quad \text{(d any divisor of m).}$$

The formulas (12.16) serve for the recursive computation of all $f_d(x)$ and thus ultimately of $f_m(x) = f(x)$. Their explicit solution depends on the Möbius function $\mu(m)$:

$\mu(m) = 0$ if m contains multiple prime factors in $\mathcal{G}$,

$$\mu(m) = \begin{cases} +1 & \text{according as m splits into an even or odd} \\ -1 & \text{number of distinct prime factors.} \end{cases}$$

One readily verifies

$$\sum_{d|m} \mu(d) = \begin{cases} 1 & \text{for } m = 1, \\ 0 & \text{for all other } m, \end{cases}$$

and thereupon obtains the solution

(12.17) $$f_m(x) = \prod_{d|m} (x^{m/d} - 1)^{\mu(d)}$$

for the recursive equations (12.15).

One could try to prove directly and without leaving the rational ground field $\mathcal{G}$, that the rational functions

$$f_1(x),\ f_2(x),\ f_3(x),\ \ldots$$

as defined by (12.17) are: (1) polynomials, and (2) polynomials irreducible in $\mathcal{G}$ (or even in any algebraic number field whose discriminant is prime to m). However, our arithmetical method yields more complete returns in a less artificial manner, although it puts at the very end the explicit (?) expression (12.17) of the polynomial $f_m(x)$ which defines the m-cyclotomic field.

The regular m-gon has an obvious geometric symmetry described by the additive group of all residues mod m. It has a hidden algebraic symmetry described by the multiplicative group of the residues prime to m (i.e., the Galois group of the cyclotomic field). However, certain geometric questions like that of constructibility by ruler and compass depend on this deep algebraic rather than on the obvious geometric symmetry.

Chapter IV

ALGEBRAIC NUMBER FIELDS

From now on the ground field will always be the rational field $\mathfrak{g}$. The finite fields k over $\mathfrak{g}$ are called algebraic number fields. In this chapter we shall deal with such features as are peculiar to them. The alphabets used are indicated by the following table:

Fields:	$\mathfrak{g} \subset$	$k \subset$	K
Numbers:	c	γ	Γ
Ideals:		$\mathfrak{a}$	$\mathfrak{A}$
Prime spots:	p	$\mathfrak{p}$	$\mathfrak{P}$

Λ is the field of all real, Ω the field of all complex numbers.

1. Lattices (old-fashioned)

In terms of a basis $\omega_1, \ldots, \omega_n$ the numbers α in k,

$$(1.1) \qquad \alpha = a_1\omega_1 + \ldots + a_n\omega_n,$$

are represented by all vectors $(a_1, \ldots, a_n)$ with <u>rational</u> components a_i in an n-dimensional vector space. Contrary to the usage in geometry only such vectors are admitted. In other words, the numbers constitute the n-dimensional $\mathfrak{g}$-vector space R which may also be looked upon as a point space with center O. The integers α form a <u>lattice</u> in that space, the word lattice meaning any set of vectors such that $\alpha \pm \beta$ are in the set if α and β are. As we shall soon realize this lattice is discrete, i.e., there is only a finite number of lattice points in any finite region, or fixing the finite region in an elementary manner: there is only a finite number of them with coördinates (components) $x_1, \ldots, x_n$ satisfying the inequalities

$$|x_i| \leq M$$

however large the natural number M. (It is clear that discreteness thus defined is independent of the choice of the basis $\omega_1, \ldots, \omega_n$.) The following study of discrete lattices in a vector space R is fundamental in many branches of mathematics.

Theorem IV 1, A. *Any discrete lattice has a (lattice) basis $\omega_1, \ldots, \omega_m$, i.e., one can choose $m \leqq n$ linearly independent lattice vectors $\omega_1, \ldots, \omega_m$ such that every vector of the lattice is a combination*

$$a_1\omega_1 + \ldots + a_m\omega_m$$

with common integers a_i as coefficients.

Proof. (1) Either the lattice consists of the vector 0 only, then our proposition holds good with m = 0; or there is at least one lattice vector $\lambda \neq 0$. λ spans a 1-dimensional manifold $\{\lambda\}$ consisting of all vectors $x\lambda$ (x rational). There is only a finite number of lattice vectors on $\{\lambda\}$ whose abscissa x satisfies the inequality $0 < x \leqq 1$. Select the one with the lowest $x = x^0$, $x^0\lambda = \omega_1$; we maintain that every lattice vector α in $\{\lambda\}$, $\alpha = x\lambda$, is an integral multiple of ω_1. Indeed one can subtract a multiple ax^0 of x^0 from x such that

$$0 \leqq x' = x - ax^0 < x^0;$$

then $\alpha - a\omega_1$ is a lattice vector of the form $x'\lambda$ with $0 \leqq x' < x^0$, and hence according to the construction of ω_1:

$$x' = 0, \qquad \alpha = a\omega_1.$$

(2) Supposing we have constructed r - 1 linearly independent lattice vectors $\omega_1, \ldots, \omega_{r-1}$ such that every lattice vector on the (r - 1)-dimensional manifold $L = \{\omega_1, \ldots, \omega_{r-1}\}$,

$$\alpha = x_1\omega_1 + \ldots + x_{r-1}\omega_{r-1},$$

has integral components $x_1, \ldots, x_{r-1}$. As under (1) there are two cases: either this exhausts all vectors of the given lattice, then our theorem is true with m = r - 1; or there is a lattice vector λ not situated in L. We consider all lattice vectors of the form

$$x_1\omega_1 + \dots + x_{r-1}\omega_{r-1} + x\lambda \quad \text{with}$$

$$0 \leqq x_1 < 1, \dots, 0 \leqq x_{r-1} < 1; \quad 0 < x \leqq 1.$$

These form a finite set containing λ. We select a vector of lowest $x = x^0$ in the set,

$$\omega_r = x_1^0\omega_1 + \dots + x_{r-1}^0\omega_{r-1} + x^0\lambda,$$

and maintain that every lattice vector α on the r-dimensional linear manifold $\{\omega_1, \dots, \omega_{r-1}, \lambda\}$ is an integral linear combination of $\omega_1, \dots, \omega_r$. Indeed one can first subtract from α a suitable multiple of ω_r and then such multiples of $\omega_1, \dots, \omega_{r-1}$ as to reduce the coefficients of the remainder

$$\alpha' = x_1'\omega_1 + \dots + x_{r-1}'\omega_{r-1} + x'\lambda$$

to the intervals

$$0 \leqq x_1' < 1, \dots, \quad 0 \leqq x_{r-1}' < 1, \quad 0 \leqq x' < x^0.$$

But then, according to the construction of ω_r, one must have first $x' = 0$ and secondly $x_1' = \dots = x_{r-1}' = 0$.

Thus the proposition is proved by induction with respect to r. The part (1) is actually superfluous; we put it in only in order to explain in simplest form the basic idea of the construction.

The discrete lattice is said to be n-dimensional, m = n, if it contains n linearly independent vectors.

Two bases

$$(1.2) \qquad \omega_1, \dots, \omega_n \;\Big|\; \omega_1^*, \dots, \omega_n^*$$

of the same n-dimensional discrete lattice are linked by relations

$$(1.3) \qquad \omega_i^* = \sum_k a_{ik}\omega_k, \qquad \omega_i = \sum_k b_{ik}\omega_k^*$$

with integral coefficients a_{ik}, b_{ik}. Since

$$|a_{ik}| \cdot |b_{ik}| = 1,$$

the determinant $|a_{ik}| = \pm 1$. A linear transformation with integral coefficients a_{ik} whose determinant $= \pm 1$ is said t be unimodular. Hence bases are connected by unimodular transformations, and the theory of lattices is essentially a theory of invariants for the group of unimodular transformations.

Our basic construction will appear here in the following form. We are given in advance a discrete n-dimensional lattice $\mathcal{L}_0$ referred to a definite basis $\tau_1, \ldots, \tau_n$, so that $\mathcal{L}_0$ consists of all vectors

$$\alpha = a_1\tau_1 + \ldots + a_n\tau_n$$

with integral components $a_1, \ldots, a_n$. We study a sublattice $\mathcal{L}$ of $\mathcal{L}_0$. Sometimes we call $\mathcal{L}_0$ the fine and $\mathcal{L}$ the coarse lattice. $\mathcal{L}$ is then surely discrete. Its n-dimensionality is secured if there exists a positive integer h such that $h\alpha$ lies in $\mathcal{L}$ for every vector α in $\mathcal{L}_0$. In applying our construction in such a way that $h\tau_r$ serves as our λ in the rth step we obtain a basis $\omega_1, \ldots, \omega_n$ for $\mathcal{L}$ of the recurrent type:

$$(1.4)\qquad \begin{aligned} \omega_1 &= c_1\tau_1, \\ \omega_2 &= c_{21}\tau_1 + c_2\tau_2, \\ &\cdots\cdots\cdots \\ \omega_n &= c_{n1}\tau_1 + \ldots + c_{n,n-1}\tau_{n-1} + c_n\tau_n \end{aligned}$$

with integral coefficients and

$$c_1 > 0, \ldots, c_n > 0.$$

The proof of our main theorem assumes a distinctly more finitistic aspect inasmuch as one chooses ω_r among the vectors

$$x_1\tau_1 + \ldots + x_{r-1}\tau_{r-1} + x\tau_r$$

with

$$\left.\begin{aligned} x_1 &= 0, 1, \ldots, c_1 - 1 \\ &\cdots\cdots\cdots \\ x_{r-1} &= 0, 1, \ldots, c_{r-1} - 1 \end{aligned}\right| \quad x = 1, \ldots, h$$

as the one with the lowest x. Minkowski calls this construction the adaptation of a coarse lattice $\mathcal{L}$ to a finer one $\mathcal{L}_0$.

Two vectors are congruent mod $\mathcal{L}$ if their difference lies in $\mathcal{L}$. Any vector of $\mathcal{L}_0$ is congruent mod $\mathcal{L}$ to one and only one of the following "residues"

$$x_1\tau_1 + \dots + x_n\tau_n, \quad 0 \leqq x_i < c_i,$$

as one readily sees by consecutive subtraction of suitable multiples of the basic vectors $\omega_n, \omega_{n-1}, \dots, \omega_1$, (1.4), of $\mathcal{L}$. Hence the density of $\mathcal{L}_0$ in $\mathcal{L}$ or the number of vectors in $\mathcal{L}_0$ which are incongruent modulo $\mathcal{L}$, amounts to

$$c_1c_2 \dots c_n.$$

This product is the determinant of the recursive substitution (1.4), $\omega = C\tau$. If we refer $\mathcal{L}$ to an arbitrary basis $\omega_1^*, \dots, \omega_n^*$ we shall have

$$\omega^* = A\omega \quad \text{and} \quad \omega^* = C^*\tau \quad \text{with} \quad C^* = AC.$$

Since A is unimodular,

$$\det C^* = \pm \det C.$$

Therefore this general result:

Theorem IV 1, B. *If the lattice $\mathcal{L}$ with the basis $\omega_1, \dots, \omega_n$ is contained in the lattice $\mathcal{L}_0$ with the basis $\tau_1, \dots, \tau_n$,*

$$\omega_i = \sum_k c_{ik}\tau_k \quad (c_{ik} \text{ integers})$$

then the density of $\mathcal{L}_0$ mod $\mathcal{L}$ is the absolute value of the determinant $|c_{ik}|$.

2. Field Basis and Basis of an Ideal

Since $\mathfrak{Q}$ is of characteristic zero, every algebraic number field k is non-degenerate.

We see in the following way that the integers in k constitute an n-dimensional discrete lattice. Start with an arbitrary basis of the field and multiply its members by a suitable rational integer such that they become integral in k. We then have a basis $\tau_1, \dots, \tau_n$ consisting of integers in k. The representation of an arbitrary integer α in terms of this basis,

$$\alpha = a_1\tau_1 + \dots + a_n\tau_n,$$

leads to

$$S(\alpha\tau_i) = \sum_k S(\tau_i\tau_k) \cdot a_k$$

and thus proves the components a_i to be of the form

$$\frac{\text{rational integers}}{h}$$

where $h = d(\tau_1 \ldots \tau_n)$ is the discriminant $|S(\tau_i\tau_k)|$ of our basis. Hence the lattice $\mathcal{L}$ of integers α is contained in the lattice $\mathcal{L}_0$ with the basis

$$\tau_1/h, \ldots, \tau_n/h$$

while conversely $h\beta$ is in $\mathcal{L}$ for any β in $\mathcal{L}_0$. We have exactly the situation as described above and thus find:

Theorem IV 2, A. *Any algebraic number field has an integral basis.*

Its discriminant is the discriminant d of the field. Two integral bases (1.2) are connected by an unimodular substitution A, (1.3), and by the law of transformation for discriminants,

$$d(\omega_1^* \ldots \omega_n^*) = (\det A)^2 \cdot d(\omega_1 \ldots \omega_n),$$

their discriminants

$$d(\omega_1^* \ldots \omega_n^*) = d(\omega_1 \ldots \omega_n) = d$$

coincide. In other words, not only is the discriminant ideal (d) uniquely determined, but also the sign of the number d is fixed unambiguously. Incidentally this is the sign for every basis, integral or not.

Let now $\mathfrak{r}$ be an ideal in k. The integers α divisible by $\mathfrak{r}$ form a sublattice $\mathcal{L}'$ of the lattice $\mathcal{L}$ of all integers. If h is a positive rational integer divisible by $\mathfrak{r}$ then $h\alpha$ is in $\mathcal{L}'$ for any α in $\mathcal{L}$. Hence the same construction again applies and we find a basis $\gamma_1, \ldots, \gamma_n$ for $\mathfrak{r}$,

$$\gamma_i = \sum_k c_{ik}\omega_k, \tag{2.1}$$

such that the set of all integers divisible by $\mathfrak{r}$ coincides with the set of all numbers of the form

$$u_1\gamma_1 + \ldots + u_n\gamma_n$$

with integral rational components u_i. (Such a basis is of a much more special character than what was formerly described as an "ideal basis" of $\mathfrak{r}$.) If one likes one can assume the transformation (2.1) to be of the recursive type (1.4).

The finite number of integers which are incongruent mod $\mathfrak{r}$ shall be called $N\mathfrak{r}$. A complete residue system will consist of $N\mathfrak{r}$ numbers. We have seen that $N\mathfrak{r}$ equals the absolute value of the determinant c_{ik},

$$N\mathfrak{r} = \left|\det c_{ik}\right|,$$

therefore

$$d(\gamma_1 \ldots \gamma_n) = (N\mathfrak{r})^2 \cdot d. \tag{2.2}$$

For later purposes we add an application of the integral field basis $\omega_1, \ldots, \omega_n$ to the enumeration of ideals which divide a given positive rational integer t. Any such ideal is of the form

$$\mathfrak{m} = (t,\alpha), \qquad \alpha = a_1\omega_1 + \ldots + a_n\omega_n. \tag{2.3}$$

Clearly (2.3) does not change if one replaces a component a_i by one $\equiv a_i \pmod t$. This leaves us with only t^n possibilities for α which may possibly lead to different $\mathfrak{m}$. Hence the number of ideal divisors of t is $\leqq t^n$.

3. Norm and Number of Residues

Theorem IV 3, A. $Nm\,\mathfrak{m} = (N\mathfrak{m})$.

There are two essentially different methods for demonstrating this important law identifying norm and number of residues of an ideal. The first method starts with the remark that the coincidence of Kronecker and Kummer degree of a prime ideal establishes our law for a prime ideal, see formula (III, 12.3). Any ideal $\mathfrak{m}$ is the product of prime ideals, $\mathfrak{m} = \mathfrak{p}\mathfrak{q} \ldots$, and

$$Nm\,\mathfrak{m} = Nm\,\mathfrak{p} \cdot Nm\,\mathfrak{q} \ldots .$$

As soon as one is able to prove the same multiplicative law for the number of residues $N\mathfrak{a}$ our theorem is established.

Theorem IV 3, B. $N(\mathfrak{a}\mathfrak{b}) = N\mathfrak{a} \cdot N\mathfrak{b}$.

Choose complete residue systems Σ, T mod $\mathfrak{a}$ and mod $\mathfrak{b}$, and a number $\alpha : \mathfrak{a}$ such that $\frac{(\alpha)}{\mathfrak{a}}$ is prime to $\mathfrak{b}$. Any pair (ξ in Σ, η in T) gives rise to an integer

$$\zeta = \xi + \alpha\eta. \tag{2.4}$$

They are all incongruent mod $\mathfrak{a}\mathfrak{b}$ since

$$\xi_1 + \alpha\eta_1 \equiv \xi_2 + \alpha\eta_2 \quad (\text{mod } \mathfrak{a}\mathfrak{b})$$

leads first to

$$\xi_1 \equiv \xi_2 \ (\mathfrak{a}), \quad \text{hence } \xi_1 = \xi_2,$$

and then to

$$\eta_1 \equiv \eta_2 \ (\mathfrak{b}), \quad \eta_1 = \eta_2.$$

Vice versa, given an integer ζ, first determine ξ in Σ by the congruence $\xi \equiv \zeta \ (\mathfrak{a})$ and then η in T by

$$\zeta - \xi \equiv \alpha\eta \quad (\mathfrak{a}\mathfrak{b}).$$

The latter congruence has a solution because $\zeta - \xi$ is divisible by the GCD $\mathfrak{a}$ of α and $\mathfrak{a}\mathfrak{b}$. Consequently every integer is congruent mod $\mathfrak{a}\mathfrak{b}$ to a number of the form (2.4). This completes the proof of Theorems IV 3, A and B.

The second method is simplest for principal ideals. If α is an integer and $\omega_1, \ldots, \omega_n$ an integral field basis then $\alpha\omega_1, \ldots, \alpha\omega_n$ is a basis of the principal ideal $\mathfrak{a} = (\alpha)$. With the equations

$$\alpha\omega_i = \sum_k a_{ki}\omega_k,$$

which yield the representing matrix $\| a_{ik} \|$ of α, one finds

$$\text{Nm } \alpha = | a_{ik} |$$

while on the other hand we have seen that this determinant $= \pm N\mathfrak{a}$.

For an arbitrary ideal $\mathfrak{a} = (\alpha_1, \ldots, \alpha_r)$ the argument works if we replace the number α by the form $\alpha_1 x_1 + \ldots + \alpha_r x_r$ which in Kronecker's theory represents the "divisor" $\mathfrak{a}$. We must make use of the following

Lemma IV 3, C. *An integral element α of $k(x, y, \ldots)$ is of the form*

$$a_1\omega_1 + \ldots + a_n\omega_n$$

where $a_1, \ldots, a_n$ are integral elements of $\mathfrak{J}(x, y, \ldots)$.

Proof. Write α as a fraction $\frac{\varphi(x,y,\ldots)}{\psi(x,y,\ldots)}$ whose numerator and denominator are polynomials with integral coefficients in k. Following the procedure in Ch. I, §9 we write

$$\mathrm{Nm}\ \psi = \psi \cdot \tau, \qquad \alpha = \frac{\varphi \cdot \tau}{\mathrm{Nm}\ \psi}\,.$$

Let N be the GCD of the integral rational coefficients of $\mathrm{Nm}\ \psi$, so that

$$\mathrm{Nm}\ \psi = N \cdot E(x, y, \ldots)$$

where $E(x,y,\ldots)$ is "primitive." The integrity of α implies that all coefficients of $\varphi \cdot \tau$ are divisible by N. Expressing all coefficients of $\varphi^* = \frac{\varphi \cdot \tau}{N}$ in terms of the integral basis $\omega_1, \ldots, \omega_n$ one gets

$$\varphi^* = f_1\omega_1 + \ldots + f_n\omega_n$$

where $f_i = f_i(x,y,\ldots)$ are polynomials with integral rational coefficients, and then

$$\alpha = \frac{f_1}{E} \cdot \omega_1 + \ldots + \frac{f_n}{E} \cdot \omega_n.$$

With this lemma at our disposal we now proceed as follows. Let $\gamma_1, \ldots, \gamma_n$ be a lattice basis of the ideal $\mathfrak{a} = (\alpha_1, \ldots, \alpha_r)$. By expressing $\alpha_1\omega_i, \ldots, \alpha_r\omega_i$ which are divisible by $\mathfrak{a}$ in terms of the basis of $\mathfrak{a}$, equations result

$$(2.5) \qquad (\alpha_1 x_1 + \ldots + \alpha_r x_r)\omega_i = \sum_k a_{ik}(x) \cdot \gamma_k$$

where the $a_{ik}(x)$ are linear forms of the indeterminates $x_1, \ldots, x_r$ with integral rational coefficients. Using the formulas (2.1) we see that the representing matrix of $\alpha_1 x_1 + \ldots + \alpha_r x_r$ with respect to the basis $\omega_1, \ldots, \omega_n$ is

$$A(x) \cdot C \quad \text{where} \quad A(x) = \| a_{ik}(x) \| , \quad C = \| c_{ik} \| .$$

By definition Nm $\mathfrak{a}$ is the content of its determinant, $|C| \cdot |A(x)|$, or Nm $\mathfrak{a}$ equals N$\mathfrak{a}$ times the GCD of the coefficients of the form det $A(x)$. All that is necessary then is to show this form to be primitive.

Because $\gamma_1, \ldots, \gamma_n$ are divisible by $\mathfrak{a}$, the elements

$$\frac{\gamma_i}{\alpha_1 x_1 + \ldots + \alpha_r x_r}$$

are integral in $k(x_1 \ldots x_r)$ and by the last lemma we have equations

$$\frac{\gamma_i}{\alpha_1 x_1 + \ldots \alpha_r x_r} = \sum_k b_{ik}(x)\omega_k$$

where $b_{ik}(x)$ are integral elements in $\mathfrak{g}(x_1 \ldots x_r)$. These equations or

$$\gamma_i = \sum_k b_{ik}(x) \cdot (\alpha_1 x_1 + \ldots + \alpha_r x_r)\omega_k$$

arise by solving (2.5) with respect to γ_i, hence

$$|a_{ik}(x)| \cdot |b_{ik}(x)| = 1$$

which shows that $|A(x)|$ is a unit in $\mathfrak{g}(x_1 \ldots x_r)$.

4. Euler's Function and Fermat's Theorem

The number of residues modulo $\mathfrak{a}$ which are relatively prime to $\mathfrak{a}$ is denoted by $\Phi(\mathfrak{a})$. The properties of this function Φ may be explored in exactly the same fashion as those of the Euler function $\varphi(a)$ which is the Φ for the ground field $\mathfrak{g}$.

Theorem IV 4, A. *If $\mathfrak{a}$ and $\mathfrak{b}$ are relatively prime then*

$$\Phi(\mathfrak{a}\mathfrak{b}) = \Phi(\mathfrak{a}) \cdot \Phi(\mathfrak{b}).$$

Proof. The congruences

$$\zeta \equiv \xi \ (\mathfrak{a}), \qquad \zeta \equiv \eta \ (\mathfrak{b})$$

establish a one-to-one correspondence

$$\zeta \rightleftarrows (\xi, \eta)$$

between the residues ζ prime to $\mathfrak{ab}$ and the pairs (ξ, η) whose two members ξ, η range over the residues prime to $\mathfrak{a}$ and $\mathfrak{b}$ respectively.

Theorem IV 4, B. *If $\mathfrak{p}$ is a prime ideal then*

$$\Phi(\mathfrak{p}^e) = (N\mathfrak{p})^e \left\{ 1 - \frac{1}{N\mathfrak{p}} \right\}.$$

Every $\mathfrak{p}$-adic integer is representable in the form

$$\rho_0 + \rho_1\pi + \rho_2\pi^2 + \ldots \qquad (\pi \text{ prime number to } \mathfrak{p})$$

with $\rho_0, \rho_1, \rho_2, \ldots$ ranging over the $P = N\mathfrak{p}$ residues mod $\mathfrak{p}$. Hence every integer α

$$\equiv \rho_0 + \rho_1\pi + \ldots + \rho_{e-1}\pi^{e-1} \ (\mathfrak{p}^e)$$

which proves again that $N(\mathfrak{p}^e) = (N\mathfrak{p})^e$. The integer α is prime to $\mathfrak{p}$ if and only if $\rho_0 \not\equiv 0 \ (\mathfrak{p})$. This reduces the number of choices for the first coefficient ρ_0 to $(N\mathfrak{p}) - 1$. Hence

$$\Phi(\mathfrak{p}^e) = (N\mathfrak{p} - 1)(N\mathfrak{p})^{e-1}.$$

If

$$\mathfrak{a} = \mathfrak{p}_1^{e_1} \mathfrak{p}_2^{e_2} \ldots$$

is the factorization of $\mathfrak{a}$ into powers of distinct prime ideals $\mathfrak{p}_1, \mathfrak{p}_2, \ldots$, the combination of our two theorems yields the explicit formula

$$\Phi(\mathfrak{a}) = N\mathfrak{a} \cdot \Pi\left(1 - \frac{1}{N\mathfrak{p}}\right),$$

the product extending to the distinct prime divisors $\mathfrak{p}$ of $\mathfrak{a}$.

Theorem IV 4, C.

$$\sum_{\mathfrak{d} | \mathfrak{n}} \Phi(\mathfrak{d}) = N\mathfrak{n}.$$

(The sum extends over all ideal divisors $\mathfrak{d}$ of $\mathfrak{n}$.)

Proof. Any residue α mod $\mathfrak{n}$ has a GCD $(\alpha, \mathfrak{n}) = \mathfrak{d}$ with $\mathfrak{n}$ which is a divisor $\mathfrak{d}$ of $\mathfrak{n}$. Vice versa, let $\mathfrak{d}$ be any ideal divisor of $\mathfrak{n}$ and choose a number $\delta : \mathfrak{d}$ such that $\frac{(\delta)}{\mathfrak{d}}$ is prime to $\frac{\mathfrak{n}}{\mathfrak{d}}$. For any integer $\alpha : \mathfrak{d}$ there is an integer ξ such that

$$\alpha \equiv \delta\xi \ (\mathfrak{n}),$$

and ξ is uniquely determined mod $\frac{\mathfrak{n}}{\mathfrak{d}}$. The common divisor $(\alpha, \mathfrak{n})$ equals $\mathfrak{d}$ if and only if ξ is prime to $\frac{\mathfrak{n}}{\mathfrak{d}}$. Hence the number of residues α with the property $(\alpha, \mathfrak{n}) = \mathfrak{d}$ equals $\Phi\left(\frac{\mathfrak{n}}{\mathfrak{d}}\right)$.

Theorem IV 4, D. *If α is prime to the ideal $\mathfrak{m}$ then*

$$\alpha^{\Phi(\mathfrak{m})} \equiv 1 \quad (\mathfrak{m}).$$

This is the analogue of Fermat's theorem and can be proved exactly like it or by the general group theoretic remark that under multiplication the relatively prime residues mod $\mathfrak{m}$ form a group of degree $\Phi(\mathfrak{m})$.

The special case

$$\alpha^{N\mathfrak{p} - 1} \equiv 1 \ (\mathfrak{p})$$

for prime ideals which implies

$$\alpha^{N\mathfrak{p}} \equiv \alpha \ (\mathfrak{p})$$

for every integer α whether $\not\equiv 0$ or $\equiv 0$ $(\mathfrak{p})$ has played a considerable rôle before. The general theory of strictly finite fields shows the existence of a primitive residue ρ such that every integer $\not\equiv 0$ $(\mathfrak{p})$ is congruent to a power of ρ mod $\mathfrak{p}$.

5. A New Viewpoint

So far we have done nothing more than to go over the general theory and add a few obvious simplifying touches for the special case we are at present concerned with. But now we shall introduce in all earnest the notion of magnitude which is peculiar to numbers and has no analogue in the general theory. From the outset we try to assign their proper place to these considerations of magnitude alongside with those of congruence or divisibility.

In the theory of quadratic forms of n variables with rational coefficients the problem of genera consists in deciding when two such (non-degenerate) forms may be transformed into one another by a (non-singular) linear transformation with rational coefficients. A necessary condition is, for every prime number p, the existence of such a transformation with p-adic coefficients, which amounts to certain congruences modulo p or modulo higher powers of p for the coefficients of the two forms. Moreover it is necessary for both forms to have the same signature; this condition is one of magnitude, requiring the existence of a transformation with real coefficients. Only both kinds of conditions together are sufficient. We may embed the rational numbers in the field of all p-adic numbers in order to describe their behaviour modulo p and all powers of p, or we may embed them into the field of all real numbers in order to describe their magnitude. Using the field of all p-adic numbers including those that represent no rational numbers facilitates operations because it has a certain completeness with respect to the purpose for which it serves. Notice that it has the power (Mächtigkeit) of the continuum! In the same manner the field of all real numbers is complete with regard to the relation of magnitude $a < b$ (as is explicitly exhibited by Hilbert's axiom of completeness). Our example of the theory of genera of quadratic forms indicates that this parallelism is of fundamental nature.

Given a rational prime number p, there corresponds to every rational number a a p-adic number $I_p(a)$, and this correspondence $a \longrightarrow I_p(a)$ is an isomorphic mapping of $\mathfrak{g}$ into the field $\mathfrak{g}(p)$ of p-adic numbers. (Formerly we were less pedantic and identified $I_p(a)$ with a.) Moreover to every rational number a there corresponds a real number $I_\infty(a)$ (which again one usually identifies with a) such that the correspondence $a \longrightarrow I_\infty(a)$ is an isomorphic mapping of $\mathfrak{g}$ into the field $\Lambda = \mathfrak{g}_\infty$ of all real numbers. Besides the finite prime spots p we introduce here an infinite prime spot ∞ thus realizing an idea which was suggested by the rational and algebraic functions of a single variable (cf. point IV in chapter III, §5). As the essential characteristic of a prime spot is considered its defining an isomorphic mapping of the given field into a field that in a certain respect is complete. Let us follow up the analogy between p and ∞ a little closer!

In the field of real numbers α we have a topology which is best described by associating with α its absolute value $|\alpha|$ which is α for positive α, is $-\alpha$ for negative α and 0 for $\alpha = 0$. The neighborhood $\mathfrak{U}(r)$ of 0 (of radius $r > 0$) consists of all numbers α satisfying the inequality $|\alpha| < r$. The neighborhood $\mathfrak{U}(r;\alpha_0)$ of any given number α_0 arises from it by translation, according to the simple remark that the real numbers form an additive group, and is hence given by $|\alpha - \alpha_0| < r$ (additive topology). In the p-adic field a number α is considered the nearer to zero the higher the power of p by which α is divisible. Any p-adic number $\alpha \neq 0$ has a definite order h,

$$\alpha = p^h \cdot \varepsilon \ (\varepsilon \text{ a unit}), \quad \alpha \sim p^h.$$

The order of a product is the sum of the orders of the factors. It appears therefore natural to introduce

$$|\alpha| = p^{-h}$$

as the absolute value of α, and of course to ascribe to 0 the absolute value 0.

$$|\alpha| < p^{-h}$$

defines the neighborhood $\mathfrak{U}_h$ of 0, and the sequence of neighborhoods

$$\ldots, \mathfrak{U}_{-2}, \mathfrak{U}_{-1}, \mathfrak{U}_0, \mathfrak{U}_1, \mathfrak{U}_2, \ldots$$

determines the additive topology of the p-adic numbers. The absolute values satisfy the relations

$$|\alpha\beta| = |\alpha| \cdot |\beta|,$$

(5.1) $$|\alpha + \beta| \leqq |\alpha| + |\beta|.$$

The latter inequality may, for finite prime spots p, be sharpened to

$$|\alpha + \beta| \leqq \max(|\alpha|, |\beta|)$$

which incidentally is equivalent to (5.1) except for p = 2. For any rational number a we define its "absolute value at p" by

$$|a|_p = |I_p(a)|$$

whether p is a finite or the infinite prime spot.

Instead of assimilating the order h to the absolute value by using h as exponent we could have gone the opposite way by taking the logarithm of the absolute value. We use this notation:

$$\log |a|_p = l_p(a).$$

Let us now pass to a finite field k over $\mathcal{R}$ as determined by an irreducible equation $f(\theta) = 0$ of degree n in $\mathcal{R}$. For a given prime ideal $\mathfrak{p}$ we chose a prime number π and then embedded the numbers α of k into the field of all π-adic numbers; or more exactly speaking, by a certain isomorphism I_π we mapped k into the field $k(\pi)$ of all π-adic numbers. If we replace π by another prime number π^* to $\mathfrak{p}$ we obtain an isomorphic mapping $I_{\pi*}$ upon the field of π^*-adic numbers. However $I_{\pi*}$ is equivalent to I_π inasmuch as we have a definite isomorphic mapping of $k(\pi^*)$ upon $k(\pi)$ effected by the substitution

$$\pi^* = \pi \cdot \varepsilon \qquad (\varepsilon \text{ a unit})$$

which enabled us to speak simply of the $\mathfrak{p}$-adic field $k(\mathfrak{p})$. Hence in speaking of the mapping $I_\mathfrak{p}$ we always mean I_π for a chosen prime number π; but we take care to introduce only concepts which are independent of the particular choice of π. As the absolute value of a non-vanishing π-adic number $\alpha \sim \pi^h$ we define

$$|\alpha| = (N\mathfrak{p})^{-h}$$

(and $|\alpha| = 0$ for $\alpha = 0$), thereby obtaining a certain additive topology in $k(\mathfrak{p})$. For any number α of k we put

$$|\alpha|_\mathfrak{p} = |I_\mathfrak{p}(\alpha)|.$$

The equation $f(\theta) = 0$ has n distinct roots within the continuum of all complex numbers. Suppose r_1 of them, $\theta_1, \ldots, \theta_{r_1}$ to be real; the others occur in pairs of conjugate complex ones:

$$\theta_1', \theta_1''; \ \ldots \ ; \theta_{r_2}', \theta_{r_2}''.$$

Sometimes we use the notation $\theta_1, \ldots, \theta_n$ for the roots in the same arrangement. The relation

$$\theta \longrightarrow \theta_1$$

determines an isomorphic mapping I_1 of k into the field Λ of all real numbers, while the relations

$$\theta \longrightarrow \theta_1', \quad \theta \longrightarrow \theta_1''$$

determine two equivalent isomorphic mappings I_1', I_1'' of k into the field Ω of all complex numbers; equivalent because they pass into each other by the automorphism $\alpha \longrightarrow \bar{\alpha}$ of Ω (transition to the conjugate-complex). We thus have r_1 isomorphic mappings $I_1, \ldots, I_{r_1}$ of k into Λ and r_2 pairs of equivalent isomorphic mappings $(I_1', I_1''), \ldots, (I_{r_2}', I_{r_2}'')$ of k into Ω. They are entirely independent of the choice of the determining number θ. It now becomes clear how to describe this situation: We have $r_1 + r_2$ infinite prime spots $\mathfrak{y}$, r_1 real and r_2 complex ones. A real finite prime spot $\mathfrak{y}$ determines an isomorphic mapping

$$\alpha \longrightarrow I_{\mathfrak{y}}(\alpha)$$

of k into the real field Λ, a complex infinite prime spot $\mathfrak{y}$ determines two equivalent isomorphic mappings

$$\alpha \longrightarrow I_{\mathfrak{y}}'(\alpha), \quad \alpha \longrightarrow I_{\mathfrak{y}}''(\alpha)$$

of k into the total complex field Ω $(I_{\mathfrak{y}}''(\alpha) = \overline{I_{\mathfrak{y}}'(\alpha)})$. When in the latter case we use the notation $I_{\mathfrak{y}}(\alpha)$ we pick out one of the two isomorphisms at random, but we introduce such concepts only as are independent of the choice.

The prime ideals are now considered as the finite prime spots. Decomposition in k,

$$(p) = \mathfrak{p}_1^{e_1} \ldots \mathfrak{p}_g^{e_g},$$

of the rational prime number p into powers of distinct prime ideals yields the g prime spots $\mathfrak{p}_1, \dots, \mathfrak{p}_g$ "lying over" the finite prime spot p of $\mathfrak{R}$ while the infinite prime spots of k are said to lie over the infinite prime spot ∞ of $\mathfrak{R}$. Their number is

$$r_1 + r_2 = r + 1 (\leqq n = r_1 + 2r_2).$$

This terminology is suggested by the analogy of algebraic functions.

The absolute value $|\alpha|$ of a real number α is again defined by

$$|\alpha| = \alpha \quad \text{if} \quad \alpha \geqq 0,$$
$$|\alpha| = -\alpha \quad \text{if} \quad \alpha \leqq 0.$$

In case of the complex field Ω we introduce, contrary to the usage in calculus,

$$|\alpha| = \alpha\bar{\alpha}$$

as the absolute value of α, thus fixing the absolute value $|\alpha|_{\mathfrak{q}}$ of a number α of k at $\mathfrak{q}$ for a real infinite prime spot by

$$|\alpha|_{\mathfrak{q}} = \pm I_{\mathfrak{q}}(\alpha) \geqq 0$$

and for a complex infinite prime spot by

$$|\alpha|_{\mathfrak{q}} = I'_{\mathfrak{q}}(\alpha) \cdot I''_{\mathfrak{q}}(\alpha).$$

(It is quite clear that the extraction of the square root would here introduce an alien irrational element.) With these conventions we obtain for any number $\alpha \neq 0$ in k:

$$|\,\mathrm{Nm}\,\alpha\,| = \prod_{\mathfrak{q}} |\alpha|_{\mathfrak{q}},$$

the product extending over the r + 1 infinite prime spots $\mathfrak{q}$.

On the other hand we derive from the decomposition of the principal ideal (α) into prime ideals,

$$(\alpha) = \prod_{\mathfrak{p}} \mathfrak{p}^h,$$

the relation

$$|\,\mathrm{Nm}\,\alpha\,| = \prod_{\mathfrak{p}} (N\mathfrak{p})^h$$

or

$$|\mathrm{Nm}\ \alpha|^{-1} = \prod_{\mathfrak{p}} |\alpha|_{\mathfrak{p}}.$$

with the product extending to all finite prime spots $\mathfrak{p}$. Hence the universal equation (Hasse)

$$\prod_{\mathfrak{p}} |\alpha|_{\mathfrak{p}} = 1 \tag{5.2}$$

in which the product runs indiscriminately over all finite and infinite prime spots. When we make use of the logarithms, writing

$$l_{\mathfrak{p}}(\alpha) = \log |\alpha|_{\mathfrak{p}}$$

for any number $\alpha \neq 0$ in k, (5.2) assumes the equivalent form

$$\sum_{\mathfrak{p}} l_{\mathfrak{p}}(\alpha) = 0. \tag{5.3}$$

We see in this equation (5.2) or (5.3) the analogue of the fact that an algebraic function has the same number of zeros and poles on a closed Riemann surface.

In older treatises algebraic number fields used to be introduced as parts of the complex continuum Ω. This is the viewpoint of analysis (and theoretical physics), but contrary to the spirit of modern algebra. We can now stake out the proper place for the "analytic" realizations $\mathfrak{g}(\theta_1), \ldots, \mathfrak{g}(\theta_n)$ of k within Λ or Ω: they amount to investigating the given field k locally at its finite prime spots. However arithmetic can not renounce studying the field locally also at every finite prime spot.

Perhaps one should not overstress the similarity between the finite and infinite prime spots; I am even willing to admit that the analogy is less close than in the theory of algebraic functions. Nevertheless the parallelism is far-reaching and in many important problems both kinds of prime spots play essentially the same rôle.

6. Minkowski's Geometric Principle

Minkowski developed a very fruitful geometric principle for arithmetical investigations in which magnitude is a decisive factor. It is concerned with a lattice of convex solids.

In a space with the n real coördinates x_i a convex solid $\mathfrak{K}$ around the origin $\mathfrak{o} = (0, \ldots, 0)$ is defined by an inequality

$$f(x_1, \ldots, x_n) \leqq 1$$

where the gauge function f has the following properties:

(i) it is continuous everywhere and >0 except at the origin;

(ii) it is homogeneous of degree 1 in the sense that

$$f(tx_1, \ldots, tx_n) = t \cdot f(x_1, \ldots, x_n) \quad \text{for every } t \geqq 0;$$

(iii) it satisfies the inequality

$$f(x_1 + y_1, \ldots, x_n + y_n) \leqq f(x_1, \ldots, x_n) + f(y_1, \ldots, y_n).$$

Simple examples are

$$\sqrt{x_1^2 + \ldots + x_n^2}$$

which defines the unit sphere and,

$$f_0(x_1, \ldots, x_n) = \max(|x_1|, \ldots, |x_n|)$$

which defines the cube

$$-1 \leqq x_1 \leqq 1, \ldots, -1 \leqq x_n \leqq 1$$

around the origin. $f(x_1, \ldots, x_n)$ assumes a positive minimum μ and maximum M on the surface of this cube,

$$f_0(x_1, \ldots, x_n) = 1;$$

hence the universal inequalities

$$\mu \cdot f_0(x) \leqq f(x) \leqq M \cdot f_0(x).$$

They prove in particular that $\mathfrak{K}$ is bounded since for all points x of $\mathfrak{K}$

$$f_0(x) \leqq 1/\mu \quad \text{or} \quad |x_i| \leqq 1/\mu \qquad (i = 1, \ldots, n).$$

A convex solid $\mathfrak{K}$ has a volume V in the Jordan sense. The inner points of $\mathfrak{K}$ are those for which $f(x) < 1$. The origin is the center of $\mathfrak{K}$ and $\mathfrak{K}$ is said to be centered (in $\mathfrak{r}$) if

$$f(-x_1, \ldots, -x_n) = f(x_1, \ldots, x_n). \tag{6.1}$$

The lattice points in our space are those whose coördinates x_i are ordinary integers. The lattice cuts the space up into the cubes or meshes

$$\mathfrak{M}(\mathfrak{a}): \quad a_1 \leqq x_1 \leqq a_1 + 1, \ldots, a_n \leqq x_n \leqq a_n + 1,$$

everyone of which is associated with a definite lattice point $\mathfrak{a} = (a_1, \ldots, a_n)$. We may shift the solid $\mathfrak{K}$ from $\mathfrak{o}$ to the lattice point $\mathfrak{a}$. In this position, $\mathfrak{K}(\mathfrak{a})$, it is defined by the inequality

$$f(x_1 - a_1, \ldots, x_n - a_n) \leqq 1.$$

Let us now assume that all the equal solids $\mathfrak{K}(\mathfrak{a})$ around the several lattice points $\mathfrak{a}$ do not overlap, or more precisely, that no inner point of $\mathfrak{K}(\mathfrak{o})$ belongs to any of the solids $\mathfrak{K}(\mathfrak{a})$ around the lattice points $\mathfrak{a} \neq \mathfrak{o}$. I then maintain that the volume V of $\mathfrak{K}$ is $\leqq 1$. This is almost trivial since there is <u>one</u> solid $\mathfrak{K}(\mathfrak{a})$ for every lattice point $\mathfrak{a}$, and V : 1 measures the portion of the space covered by the lattice of solids $\mathfrak{K}(\mathfrak{a})$. There are several ways of converting this argument into a strict proof. Here is a geometrically very simple one which puts it as a jigsaw puzzle.

The meshes of our lattice cut $\mathfrak{K} = \mathfrak{K}(\mathfrak{o})$ up into a number of convex pieces. One of these pieces is the intersection $\mathfrak{K} \cap \mathfrak{M}(\mathfrak{a})$; since $\mathfrak{K}$ is bounded, there is indeed only a finite number of such pieces. The sum total of their volumes of course is V. On the other hand we watch how all the solids $\mathfrak{K}(\mathfrak{a})$ intersect with the one mesh $\mathfrak{M} = \mathfrak{M}(\mathfrak{o})$. Again we obtain a finite number of non-overlapping pieces of $\mathfrak{M}$ whose volumes therefore must have a sum total $\leqq 1$. However these pieces are exactly the same as before, and by putting them together to form $\mathfrak{K}$ we prove $V \leqq 1$. I give the secret of the solution of the puzzle: by imparting the translation $\mathfrak{a}$ to the piece $\mathfrak{M} \cap \mathfrak{K}(-\mathfrak{a})$ of $\mathfrak{M}$ one gets $\mathfrak{K} \cap \mathfrak{M}(\mathfrak{a})$.

We are now prepared to prove Minkowski's fundamental theorem:

If the centered convex solid $\mathfrak{K}$ around the origin has a volume $> 2^n$ then it contains at least one lattice point $\neq \mathfrak{o}$ in its interior.

Indeed the solid $\frac{1}{2}\mathfrak{K}$ arising from $\mathfrak{K}$ by contraction in the scale 1 : 2 has a volume > 1, hence it must have an inner point $\mathfrak{c}$ in common with one of the solids $\frac{1}{2}\mathfrak{K}(\mathfrak{a})$ around the lattice points $\mathfrak{a} \neq \mathfrak{o}$; or

$$f(\mathfrak{u}) < \frac{1}{2}, \; f(\mathfrak{u} - \mathfrak{v}) \leqq \frac{1}{2}.$$

By (6.1) the second inequality may be replaced by

$$f(\mathfrak{v} - \mathfrak{u}) \leqq \frac{1}{2}$$

and the condition (iii) then yields

$$f(\mathfrak{v}) \leqq f(\mathfrak{u}) + f(\mathfrak{v} - \mathfrak{u}) < \frac{1}{2} + \frac{1}{2} = 1.$$

We need not bother about any set-theoretic subtleties because we are going to apply this principle to a very elementary type of solids only which we call elliptic cylinders. We consider our space as an affine space. The inequalities

$$|x_1| < a_1, \ldots, |x_n| < a_n$$

then describe what the topologist would call direct (affine) product of n segments, but what is commonly termed a parallelotope. $a_1, \ldots, a_n$ are given positive numbers. Its volume is $2^n \cdot a_1 \ldots a_n$. Let us sift out r_2 pairs of the coördinates, so that they are now denoted by

$$x_1, \ldots, x_{r_1}; \; u_1, v_1, \ldots, u_{r_2}, v_{r_2} \quad (r_1 + 2r_2 = n)$$

and consider the inequalities

$$|x_1| < a_1, \ldots, |x_{r_1}| < a_{r_1}; \; u_1^2 + v_1^2 < b_1, \ldots, u_{r_2}^2 + v_{r_2}^2 < b_{r_2}.$$

The solid defined by them is the direct (affine) product of r_1 segments and r_2 ellipses, and since in the lowest case $r_1 = 1$, $r_2 = 1$, $n = 3$, this is an elliptic cylinder we shall in general refer to it by that name. Its volume amounts to

$$2^{r_1}\pi^{r_2} a_1 \ldots a_{r_1} b_1 \ldots b_{r_2}.$$

If we introduce the conjugate complex coördinates

$$u_\mu + iv_\mu = x'_\mu, \qquad u_\mu - iv_\mu = x''_\mu$$

and, as formerly advocated, interpret $|\alpha|$ as $\alpha\bar{\alpha}$ in the complex case, we may describe our cylinder by the inequalities

$$|x_1| < a_1, \ldots, |x_{r_1}| < a_{r_1};$$

$$|x_1'| = |x_1''| < b_1, \ldots, |x_{r_2}'| = |x_{r_2}''| < b_{r_2}.$$

By an arbitrary linear transformation with real coefficients we pass to another affine coördinate system and then obtain the following special case of Minkowski's theorem:

Theorem IV 6, A. *Given r_1 real and r_2 pairs of conjugate complex linear forms of the $n = r_1 + 2r_2$ real variables x_i:*

$$(6.2)\quad \begin{aligned} y_\lambda &= a_{\lambda 1}x_1 + \cdots + a_{\lambda n}x_n && (\lambda = 1, \ldots, r_1), \\ \left\{ \begin{array}{l} y_\mu' = a_{\mu 1}'x_1 + \cdots + a_{\mu n}'x_n \\ y_\mu'' = a_{\mu 1}''x_1 + \cdots + a_{\mu n}''x_n \end{array} \right. && && (\mu = 1, \ldots, r_2) \end{aligned}$$

and $r_1 + r_2$ positive numbers a_λ, b_μ. We assume the determinant of the n forms y to be $\neq 0$ and designate its absolute value by D. Then the inequalities

$$(6.3)\qquad |y_\lambda| < a_\lambda, \quad |y_\mu'| = |y_\mu''| < b_\mu$$

describe an elliptic cylinder of volume

$$2^{r_1}(2\pi)^{r_2}a_1 \ldots a_{r_1}b_1 \ldots b_{r_2}/D.$$

Hence if

$$a_1 \ldots a_{r_1}b_1 \ldots b_{r_2} > \left(\frac{2}{\pi}\right)^{r_2} D$$

the inequalities (6.3) have an integral solution

$$(x_1, \ldots, x_n) \neq (0, \ldots, 0).$$

Notice that in (6.3) we have the sign < throughout and not $\leq$. The gauge function of our elliptic cylinder is the maximum of the $r_1 + r_2$ quantities

$$\frac{|y_\lambda|}{a_\lambda}; \quad \frac{|y_\mu'|^{\frac{1}{2}}}{b_\mu^{\frac{1}{2}}} = \frac{|y_\mu''|^{\frac{1}{2}}}{b_\mu^{\frac{1}{2}}}.$$

How much can one say in the limiting case

$$a_1 \ldots a_{r_1}b_1 \ldots b_{r_2} = \left(\frac{2}{\pi}\right)^{r_2}D \ ?$$

Single out one of the inequalities (6.3), say the first one, and replace a_1 by $2a_1$. Then they have a non-trivial integral solution, but of course only a finite number. Take the one for which $|y_1|$ assumes its lowest value a. Since the inequalities

$$|y_1| < a, \ |y_2| < a_2, \ldots, \ |y_{r_1}| < a_{r_1}, \ |y'_\mu| = |y''_\mu| < b_\mu$$

now have no integral solution $\neq 0$, the inequality

$$aa_2 \ldots a_{r_1} b_1 \ldots b_{r_2} > \left(\frac{2}{\pi}\right)^{r_2} D$$

would contradict Minkowski's theorem and thus we have

$$aa_2 \ldots a_{r_1} b_1 \ldots b_{r_2} \leqq \left(\frac{2}{\pi}\right)^{r_2} D \quad \text{or} \quad a \leqq a_1.$$

In other words we get this

Supplement: *If*

$$a_1 \ldots a_{r_1} b_1 \ldots b_{r_2} = \left(\frac{2}{\pi}\right)^{r_2} D$$

then the inequalities

$$|y_\lambda| \leqq a_\lambda, \quad |y'_\mu| = |y''_\mu| \leqq b_\mu$$

have a non-trivial integral solution such that the sign < prevails in all of them but one which has been singled out in advance.

7. A Fundamental Inequality and Its Consequences: Existence of Ramification Ideals, Classes of Ideals

Let our field k over $\mathfrak{R}$ again have r_1 real and r_2 complex infinite prime spots and d be its discriminant. We put

$$\delta = \left(\frac{2}{\pi}\right)^{r_2} \cdot \sqrt{|d|}.$$

For any ideal $\mathfrak{a}$ in k we introduce an integral basis $\iota_1, \ldots, \iota_n$ such that the linear form

$$\iota_1 x_1 + \ldots + \iota_n x_n \tag{7.1}$$

yields all numbers α divisible by $\mathfrak{m}$ in substituting for x_i arbitrary rational integers. We apply the result of the last section to the n conjugates of (7.1) in Ω,

$$y_\lambda; \quad y'_\mu, \; y''_\mu.$$

The square of their determinant equals $d \cdot (N\mathfrak{m})^2$. In preceding our proposition IV 6, A by the lemma that the inequalities (6.3) have only a finite number of solutions whatever the positive values a_λ, b_μ we then get these facts:

Lemma IV 7, A. *Let a positive number $\upsilon_\mathfrak{p}$ be assigned to each infinite prime spot $\mathfrak{p}$. There is only a finite number of numbers α in k which are divisible by the given ideal $\mathfrak{m}$ and satisfy the inequalities*

$$|\alpha|_\mathfrak{p} \leqq \upsilon_\mathfrak{p}. \tag{7.2}$$

Theorem IV 7, B. *If*

$$\prod_\mathfrak{p} \upsilon_\mathfrak{p} \geqq \delta \cdot N\mathfrak{m} \tag{7.3}$$

then the inequalities (7.2) have a solution $\alpha \neq 0$ in k which is divisible by $\mathfrak{m}$.

Supplement. *If*

$$\prod_\mathfrak{p} \upsilon_\mathfrak{p} > \delta \cdot N\mathfrak{m}$$

one may even replace the sign $\leqq$ by $<$ in all inequalities (7.2); if (7.3) holds, in all inequalities save one.

It should be observed that these facts hold for a fractional no less than for an integral ideal $\mathfrak{m}$. Indeed a fractional ideal

$$\mathfrak{m} = \mathfrak{b}/\mathfrak{c} \qquad (\mathfrak{b}, \mathfrak{c} \text{ integral ideals})$$

can be written as $\mathfrak{m} = \frac{\mathfrak{m}^*}{c}$ where c is a positive rational integer and $\mathfrak{m}^*$ an integral ideal. If $\iota_1^*, \ldots, \iota_n^*$ is a lattice basis for $\mathfrak{m}^*$ then

$$\iota_1 = \iota_1^*/c, \ldots, \iota_n = \iota_n^*/c$$

plays the same rôle for $\mathfrak{a}$, and $N\mathfrak{a}$ is defined as

$$N\mathfrak{b}/N\mathfrak{c} = N\mathfrak{a}^*/c^n.$$

In decomposing $\mathfrak{a}$ into powers of distinct prime ideals,

$$\mathfrak{a} = \mathfrak{p}_1^{h_1} \cdots \mathfrak{p}_t^{h_t} \tag{7.4}$$

we need not, therefore, assume the exponents $h_1, \ldots, h_t$ to be $\geqq 0$. But under all circumstances

$$N\mathfrak{a} = (N\mathfrak{p}_1)^{h_1} \ldots (N\mathfrak{p}_t)^{h_t}.$$

If we multiply all the inequalities (7.2), choosing for $v_{\mathfrak{p}}$ any positive numbers whose product equals $\delta \cdot N\mathfrak{a}$, we find:

Theorem IV 7, C. *For every (integral or fractional) ideal $\mathfrak{a}$, there exists a number $\alpha \neq 0$ in k which is divisible by $\mathfrak{a}$ such that*

$$|Nm\ \alpha| \leqq \delta \cdot N\mathfrak{a}.$$

Supplement. *The sign $\leqq$ may be replaced by $<$ except when there is only one infinite prime spot, i.e., except for the rational ground field and the imaginary quadratic field.*

In writing

$$(\alpha) = \mathfrak{a}.\mathfrak{b},$$

$\mathfrak{b}$ is an integral ideal satisfying the inequality

$$N\mathfrak{b} \leqq \delta.$$

Hence the equivalent statement:

Theorem IV 7, D. *Given an (integral or fractional) ideal $\mathfrak{a}$, there always exists an integral ideal $\mathfrak{b}$ such that*

$$\mathfrak{a}\mathfrak{b} \text{ is principal and } N\mathfrak{b} \leqq \delta.$$

If we specialize Theorem IV 7, C and its supplement

for $\mathfrak{a} = (1)$ we see that there exists an integer $\alpha \neq 0$ such that

$$|\mathrm{Nm}\,\alpha|^2 < |d|,$$

except for $k = \mathcal{Q}$. This inequality implies $|d| > 1$, or divisibility of d by at least one prime number. In other words:

Theorem IV 7, E. *Every field k over $\mathcal{Q}$, except $\mathcal{Q}$ itself, has finite ramification prime spots.*

The corresponding proposition in the theory of algebraic functions asserts that a Riemann surface extending without ramification over the complex z-plane (excluding $z = \infty$) necessarily consists of one sheet only. This is a fundamental topological principle called by Weierstress the principle of monodromy. (The z-plane may here be replaced by a simply connected surface.) It is remarkable how different the methods are by which one proves the existence of ramification spots in the cases of algebraic numbers and functions. By more refined applications of Minkowski's geometric principle one is able to derive much stronger inequalities.

The fractional ideals form a group of which the principal ideals form a subgroup. Two ideals $\mathfrak{a}$, $\mathfrak{b}$ in the same coset modulo this subgroup are said to belong to the same class or to be equivalent, $\mathfrak{a} \simeq \mathfrak{b}$. This is the case if and only if $\mathfrak{b}$ arises from $\mathfrak{a}$ by multiplication with a number $\alpha \neq 0$. We state the fundamental fact that the subgroup of principal ideals is of finite index H in the group of all ideals, or that the number H of classes is finite. Indeed if one replaces the ideal $\mathfrak{a}$ in Theorem IV 7, D by $\mathfrak{a}^{-1}$ one arrives at the following proposition:

Theorem IV 7, F. *In every class there exists an integral ideal $\mathfrak{b}$ whose norm $N\mathfrak{b} \leqq \delta$.*

Such an ideal is a divisor of one of the numbers

$$1, 2, \ldots, [\delta].$$

In making use of a previous estimate according to which the positive rational integer t has at most t^n ideal divisors we find that the class number

$$H \leqq 1^n + 2^n + \ldots + [\delta]^n.$$

If the class number = 1, every ideal is a principal ideal and numbers are the only divisors.

It is of course easy to describe the distribution of ideals in classes without resorting to fractional ideals. Two integral ideals $\mathfrak{a}$, $\mathfrak{b}$ are equivalent if there exist two integers α, $\beta \neq 0$ such that $\beta\mathfrak{a} = \alpha\mathfrak{b}$. A certain analogy to chemistry is stressed by saying molecule for integer or principal ideal and radical for (integral) ideal. In a molecule $(\alpha) = \mathfrak{a}\mathfrak{a}'$ the radical $\mathfrak{a}$ can be exchanged for the radical $\mathfrak{b}$ if $\mathfrak{b}\mathfrak{a}' = (\beta)$ is also a complete molecule. It is a fact that then $\mathfrak{a}$ may be exchanged for $\mathfrak{b}$ in any molecule of which $\mathfrak{a}$ is a part, and we speak of equivalence of the two radicals, $\mathfrak{a} \simeq \mathfrak{b}$. Indeed if $(\alpha^*) = \mathfrak{a}\mathfrak{a}^*$ then $\dfrac{\alpha^*\beta}{\alpha} = \beta^*$ is an integer and $(\beta^*) = \mathfrak{b}\mathfrak{a}^*$.

The fundamental theorem IV 7, B with its accompanying lemma and supplement can be stated in still another way. The fact that α is divisible by $\mathfrak{a}$, (7.4), may be expressed by the inequalities

$$|\alpha|_{\mathfrak{p}_i} \leqq v_{\mathfrak{p}_i} = (N\mathfrak{p}_i)^{-h_i} \quad (i = 1, \ldots, t),$$

$$|\alpha|_{\mathfrak{p}} \leqq 1 \quad \text{for all finite prime spots } \mathfrak{p} \neq \mathfrak{p}_1, \ldots, \mathfrak{p}_t.$$

Again we write

$$(7.5) \qquad (\alpha) = \mathfrak{p}_1^{h_1} \cdots \mathfrak{p}_t^{h_t} \cdot \mathfrak{b},$$

so that

$$|Nm\ \alpha| = (N\mathfrak{p}_1)^{h_1} \cdots (N\mathfrak{p}_t)^{h_t} \cdot N\mathfrak{b} \quad \text{or}$$

$$N\mathfrak{b} = |Nm\ \alpha| \cdot v_{\mathfrak{p}_1} \cdots v_{\mathfrak{p}_t}.$$

Hence the following formulation:

Lemma IV 7, G. *Let a positive number $v_{\mathfrak{p}}$ be assigned to every prime spot $\mathfrak{p}$, with these restrictions:*

(i) $v_{\mathfrak{p}} = 1$ for almost all $\mathfrak{p}$ (i.e., allowing but for a finite number of exceptions).

(ii) $v_{\mathfrak{p}}$ is of the form $(N\mathfrak{p})^{-h}$ with an integral exponent $-h$ for every finite prime spot.

Then there exist only a finite number of numbers α in k satisfying all the inequalities

$$(7.6) \qquad |\alpha|_{\mathfrak{p}} \leqq v_{\mathfrak{p}}.$$

If one forms the ideal

$$\mathfrak{n} = \prod_{\mathfrak{p}} \mathfrak{p}^{h} \qquad (\mathfrak{p} \text{ all finite prime spots})$$

and sets

$$(7.7) \qquad (\alpha) = \mathfrak{n}\mathfrak{b}$$

for a number α satisfying (7.6), then $\mathfrak{b}$ is an integral ideal for which

$$(7.8) \qquad N\mathfrak{b} \leqq \prod_{\mathfrak{p}} v_{\mathfrak{p}} \qquad (\mathfrak{p} \text{ all prime spots}).$$

Theorem IV 7, H. *If the numbers $v_{\mathfrak{p}}$ introduced in the previous lemma and subject to the conditions (i) and (ii) thereof have a product*

$$(7.9) \qquad \prod_{\mathfrak{p}} v_{\mathfrak{p}} \geqq \delta,$$

then there certainly exists a number $\alpha \neq 0$ in k satisfying all the inequalities (7.6).

Supplement. *According as the sign $>$ or $=$ holds in (7.9) one can require the sign $<$ to hold in (7.6) for all infinite prime spots or for all such prime spots save one (singled out in advance).*

8. The Dirichlet-Minkowski-Hasse-Chevalley Construction of Units

What we have called units are numbers η in k which are units at every finite prime spot or which satisfy the relation

$$|\eta|_{\mathfrak{p}} = 1$$

for all prime spots $\mathfrak{p}$ except the infinite ones. Following Hasse we replace the set S_∞ of the infinite prime spots by any finite set $S \supset S_\infty$ of prime spots and study the numbers η which are units at all prime spots outside S:

$$|\eta|_{\mathfrak{p}} = 1 \quad \text{if } \mathfrak{p} \text{ not in S.}$$

One could conveniently call them the units of k relative to S; they form a group U_S under multiplication. We single out one of the m + 1 prime spots in S, say $\mathfrak{p}_0$, and then show:

Theorem IV 8, A. *There exists an element η of U_S such that*

(8.1) $|\eta|_{\mathfrak{p}_0} > 1$ *and* $|\eta|_{\mathfrak{p}} < 1$ *for all* $\mathfrak{p} \neq \mathfrak{p}_0$ *in* S.

Choosing for $\mathfrak{p}_0$ each of the m + 1 prime spots $\mathfrak{p}_0, \mathfrak{p}_1, \ldots, \mathfrak{p}_m$ of S in turn, we construct m + 1 elements $\eta_0, \eta_1, \ldots, \eta_m$ in U_S such that

$$|\eta_i|_{\mathfrak{p}_i} > 1, \qquad |\eta_i|_{\mathfrak{p}_k} < 1 \quad (i \neq k)$$
$$(i,\ k = 0,\ 1, \ldots, m).$$

The construction takes place within the totality k_S^* of all numbers $\alpha \neq 0$ which are integers at the prime spots outside S. Our last theorem IV 7, H was concerned with these numbers. Throughout this section $\mathfrak{p}$ and $\mathfrak{q}$ denote prime spots in S, and $\mathfrak{p}_0$ is one of them. If there is associated with each of the m prime spots $\mathfrak{p} \neq \mathfrak{p}_0$ in S a positive number $v_{\mathfrak{p}}$, of the form $(N\mathfrak{p})^{-h}$ in case $\mathfrak{p}$ is finite, we construct a corresponding number α in k_S^* as follows.

(i) If $\mathfrak{p}_0$ is infinite we set

$$v_{\mathfrak{p}_0} = \delta / \prod_{\mathfrak{p}} v_{\mathfrak{p}} \qquad (\mathfrak{p} \neq \mathfrak{p}_0 \text{ in S}).$$

By Theorem IV 7, H and its supplement there exists a number α in k_S^* such that

$$(8.2) \qquad \begin{gathered} |\alpha|_{\mathfrak{p}_0} \leqq v_{\mathfrak{p}_0}, \\ |\alpha|_{\mathfrak{q}} < v_{\mathfrak{q}}, \qquad |\alpha|_{\mathfrak{p}} \leqq v_{\mathfrak{p}} \end{gathered}$$

where $\mathfrak{p}$ runs over all finite and $\mathfrak{q}$ over all infinite prime spots $\neq \mathfrak{p}_0$ in S. With the notation used in the lemma one has

$$N\mathfrak{b} \leqq \delta.$$

(ii) If $\mathfrak{y}_0$ is finite we again form

$$v^* = \delta / \prod_{\mathfrak{y}} v_{\mathfrak{y}} \qquad (\mathfrak{y} \neq \mathfrak{y}_0)$$

and define a $v_{\mathfrak{y}_0}$ of the form $(N\mathfrak{y}_0)^{-h_0}$ by the inequalities

$$v^* < v_{\mathfrak{y}_0} \leqq v^* \cdot N\mathfrak{y}_0.$$

Since the product

$$\prod_{\mathfrak{y}} v_{\mathfrak{y}} \qquad (\mathfrak{y} \text{ in } S) \tag{8.3}$$

now is actually $> \delta$ there again exists a number α in k_S^* satisfying the relations (8.2). The product (8.3) now is $\leqq \delta \cdot N\mathfrak{y}_0$. So we find in both cases (i) and (ii)

$$N\mathfrak{b} \leqq \delta_0$$

with $\delta_0 = \delta$ for infinite and $\delta_0 = \delta \cdot N\mathfrak{y}_0$ for finite $\mathfrak{y}_0$.

There is only a finite number of elements α in k_S^* satisfying the relations (8.2). For any $\mathfrak{y} \neq \mathfrak{y}_0$ in S we denote by $m_{\mathfrak{y}}$ the positive minimum of $|\alpha|_{\mathfrak{y}}$ for these elements and then set

$$v'_{\mathfrak{y}} = m_{\mathfrak{y}} \qquad (\mathfrak{y} \text{ infinite}),$$

$$v'_{\mathfrak{y}} = m_{\mathfrak{y}} \cdot (N\mathfrak{y})^{-1} \qquad (\mathfrak{y} \text{ finite}).$$

Notice that $m_{\mathfrak{y}}$ is of the form $(N\mathfrak{y})^{-h}$ for finite $\mathfrak{y}$. By means of these numbers $v'_{\mathfrak{y}}$ we construct a number α' in the same manner as α has before been constructed by means of $v_{\mathfrak{y}}$. As $|\alpha'|_{\mathfrak{y}} < m_{\mathfrak{y}}$ for $\mathfrak{y} \neq \mathfrak{y}_0$, namely $\leqq v'_{\mathfrak{y}} < m_{\mathfrak{y}}$ for the finite and $< v'_{\mathfrak{y}} = m_{\mathfrak{y}}$ for the infinite $\mathfrak{y} \neq \mathfrak{y}_0$, one has a fortiori

$$|\alpha'|_{\mathfrak{y}} < |\alpha|_{\mathfrak{y}} \qquad \text{for } \mathfrak{y} \neq \mathfrak{y}_0.$$

According to construction there is no element α in k_S^* such that

$$|\alpha|_{\mathfrak{y}_0} \leqq v_{\mathfrak{y}_0}, \qquad |\alpha|_{\mathfrak{y}} < m_{\mathfrak{y}} \qquad (\text{for all } \mathfrak{y} \neq \mathfrak{y}_0).$$

Hence $|\alpha'|_{\mathfrak{y}_0} > v_{\mathfrak{y}_0}$ or a fortiori

$$|\alpha'|_{\mathfrak{y}_0} > |\alpha|_{\mathfrak{y}_0}.$$

Just as we passed from α to α' we may pass from α' to α'', and so on, and we thus ascertain an infinite sequence $\alpha, \alpha', \alpha'', \ldots$ of elements in k_S^* such that the series

$$|\alpha|_{\mathfrak{p}_0}, \ |\alpha'|_{\mathfrak{p}_0}, \ |\alpha''|_{\mathfrak{p}_0}, \ \ldots$$

is increasing while the other m series

$$|\alpha|_{\mathfrak{p}}, \ |\alpha'|_{\mathfrak{p}}, \ |\alpha''|_{\mathfrak{p}}, \ \ldots$$

which correspond to the prime spots $\mathfrak{p} \neq \mathfrak{p}_0$ in S are decreasing. Moreover the integral ideals

$$\mathfrak{b}, \ \mathfrak{b}', \ \mathfrak{b}'', \ \ldots$$

analogous to $\mathfrak{b}$ all satisfy the inequality

$$N\mathfrak{b} \leqq \delta_0. \tag{8.4}$$

One will remember the equations (7.7), (7.4):

$$(\alpha) = \prod_{\mathfrak{p}} \mathfrak{p}^h \cdot \mathfrak{b} \qquad (\mathfrak{p} \text{ finite prime spots in S}).$$

According to (8.4) there is only a finite number of different ideals in our infinite sequence $\mathfrak{b}, \mathfrak{b}', \mathfrak{b}'', \ldots$. After at most

$$1^n + 2^n + \ldots + [\delta_0]^n$$

steps we must have encountered two equal $\mathfrak{b}$. If for instance $\mathfrak{b}^{(5)} = \mathfrak{b}^{(13)}$ then

$$\eta = \alpha^{(13)}/\alpha^{(5)}$$

contains only the prime ideals of S with an exponent $\neq 0$ and is therefore an element of U_S. It satisfies the desired inequalities (8.1).

9. The Structure of the Group of Units

With each element η of the group U_S we may associate its m + 1 logarithms

$$l_i(\eta) = \log |\eta|_{\mathfrak{p}_i} \qquad (i = 0, 1, \ldots, m)$$

or the "vector of η" with the components

$$z_i = l_i(\eta) \qquad (i = 0, 1, \ldots, m).$$

In truth, because of the relation

$$(9.1) \qquad l_0(\eta) + l_1(\eta) + \ldots + l_m(\eta) = 0$$

this is a vector in the m-dimensional subspace

$$z_0 + z_1 + \ldots + z_m = 0.$$

Transition from η to its vector changes the multiplicative group of the η into the additive group or lattice of their vectors. We maintain that it results in an m-dimensional discrete lattice and therefore possesses a lattice basis consisting of m members.

On account of (9.1) one may throw away one of the logarithms, say l_0. I first prove the linear independence of the vectors

$$(9.2) \qquad \begin{matrix} (l_1(\eta_1), \ldots, l_m(\eta_1)), \\ \cdot\ \cdot\ \cdot\ \cdot\ \cdot\ \cdot\ \cdot\ \cdot\ \cdot\ \cdot\ \cdot \\ (l_1(\eta_m), \ldots, l_m(\eta_m)) \end{matrix}$$

which correspond to the relative units $\eta_1, \ldots, \eta_m$ as constructed in the previous section (omitting η_0). Indeed denote the matrix (9.2) by $\| c_{ik} \|$. One has for instance

$$c_{11} > 0; \qquad c_{12} < 0, \ldots, c_{1m} < 0$$

and

$$c_{11} + \ldots + c_{1m} = -l_0(\eta_1) > 0.$$

Hence the conditions

$$\begin{matrix} |c_{11}| - |c_{12}| - \ldots - |c_{1m}| > 0, \\ -|c_{21}| + |c_{22}| - \ldots - |c_{2m}| > 0, \\ \cdot\ \cdot\ \cdot\ \cdot\ \cdot\ \cdot\ \cdot\ \cdot\ \cdot\ \cdot\ \cdot\ \cdot\ \cdot\ \cdot\ \cdot \\ -|c_{m1}| - |c_{m2}| - \ldots + |c_{mm}| > 0. \end{matrix}$$

I maintain that under these conditions the homogeneous equations

$$(9.3) \qquad \sum_k c_{ik} z_k = 0 \qquad (i, k = 1, \ldots, m)$$

have no other real solution than $z_1 = \ldots = z_m = 0$. Indeed if they had a non-trivial solution z_i we should pick out the greatest of the absolute values $|z_i|$, say $|z_1|$; we may then assume $|z_1| = 1$ and should find

$$\left|\sum_k c_{1k} z_k\right| \geqq |c_{11}| - |c_{12}| - \ldots - |c_{1m}| > 0,$$

contrary to the supposed equations (9.3).

Once it is certain that the vectors of the relative units η form an m-dimensional lattice one can write for every element η of U_S:

$$(9.4) \qquad l_i(\eta) = \sum_k l_i(\eta_k) \cdot z_k \qquad (i, k = 1, \ldots, m).$$

Because of (9.1) these equations imply the further one

$$(9.5) \qquad l_0(\eta) = \sum_k l_0(\eta_k) \cdot z_k.$$

There are only a finite number of elements η for which

$$(9.6) \qquad 0 \leqq z_1 < 1, \ldots, 0 \leqq z_m < 1,$$

and thus the lattice is discrete. In fact the inequalities (9.6) imply

$$l_1(\eta) \leqq l_1(\eta_1), \ldots, l_m(\eta) \leqq l_m(\eta_m)$$

and in view of (9.5) also

$$l_0(\eta) \leqq 0;$$

or

$$|\eta|_{\mathfrak{g}_0} \leqq 1, \quad |\eta|_{\mathfrak{g}_1} \leqq |\eta_1|_{\mathfrak{g}_1}, \ldots, |\eta|_{\mathfrak{g}_m} \leqq |\eta_m|_{\mathfrak{g}_m},$$

and we know that these inequalities, together with

$$|\eta|_{\mathfrak{g}} \leqq 1 \quad \text{for all } \mathfrak{g} \text{ not in } S,$$

have only a finite number of solutions.

The lattice basis which can now be constructed by the general device explained in §1 yields m relative units $\varepsilon_1, \ldots, \varepsilon_m$ such that every η is expressible in one and only one way as

$$(9.7) \qquad \eta = \zeta \cdot \varepsilon_1^{u_1} \ldots \varepsilon_m^{u_m}$$

where $u_1, \ldots, u_m$ are integral exponents and ζ is a unit at every prime spot whatsoever,

$$|\zeta|_{\mathfrak{p}} = 1 \quad \text{for all } \mathfrak{p}.$$

The ζ's of this property form a group which remains to be determined. ζ is an integer whose n conjugates $\zeta_1, \ldots, \zeta_n$ in Ω are of absolute value 1. Consequently the field equation for ζ,

$$x^n + c_1 x^{n-1} + \ldots + c_n = (x - \zeta_1) \ldots (x - \zeta_n)$$

has rational integral coefficients $c_1, \ldots, c_n$ which satisfy the inequalities

$$|c_1| \leqq n, \quad |c_2| \leqq \frac{n(n-1)}{1 \cdot 2}, \ldots, |c_n| \leqq 1.$$

This allows only a finite number of possibilities, consequently the ζ form a group of some finite degree w. The equation $\zeta^w = 1$ then holds for every ζ, and the result is that the group in question consists of all (w) wth roots of unity. Incidentally if there is at least one real prime spot $\mathfrak{q}$ then $|I_{\mathfrak{q}}(\zeta)| = 1$ implies $I_{\mathfrak{q}}(\zeta) = \pm 1$, hence $\zeta = \pm 1$, so that in that case w is necessarily 2.

Theorem IV 9, A. *The relative units η have a basis $\varepsilon_1, \varepsilon_2, \ldots, \varepsilon_m$ such that every η is uniquely expressible in the form (9.7) where ζ is a root of unity.*

In other words, the group U_S is the direct product of a finite cyclic group of order w and m infinite cyclic groups.

Up to its sign the determinant

$$(9.8) \qquad \begin{vmatrix} l_1(\varepsilon_1), \ldots, l_m(\varepsilon_1) \\ \cdot \cdot \cdot \cdot \cdot \cdot \cdot \cdot \cdot \cdot \\ l_1(\varepsilon_m), \ldots, l_m(\varepsilon_m) \end{vmatrix}$$

is independent of the choice of the basis $\varepsilon_1, \ldots, \varepsilon_m$. Its absolute value is called the regulator R (for the set S). Because of the relation (9.1) the absolute value of (9.8) is the same for each of the m + 1 minors of the full matrix

$$\begin{Vmatrix} l_0(\varepsilon_1), \ldots, l_m(\varepsilon_1) \\ \cdot \cdot \cdot \cdot \cdot \cdot \cdot \cdot \cdot \cdot \\ l_0(\varepsilon_m), \ldots, l_m(\varepsilon_m) \end{Vmatrix}$$

so that the regulator does not depend on which of the m + 1 logarithms is dropped.

The specialization for the lowest case $S = S_\infty$ is immediate. Then U_S is the group of the units of k. The regulator for this set S_∞ is simply called the regulator (of k).

v.d. Waerden has developed a method by which one handles the multiplicative group of the η directly without introducing their logarithms.

10. Finite Abelian Groups and Their Characters

The (fractional) ideals constitute an Abelian group of which the principal ideals form an (invariant) subgroup of finite index H. The factor group is the group of classes of ideals which is of finite order H. The unit element of this group is the principal class containing all principal ideals. The classes form a group because

$$\mathfrak{a}_1 \simeq \mathfrak{a}_2, \quad \mathfrak{b}_1 \simeq \mathfrak{b}_2 \quad \text{imply} \quad \mathfrak{a}_1\mathfrak{a}_2 \simeq \mathfrak{b}_1\mathfrak{b}_2.$$

The class group is probably the most important arithmetic characteristic of a field.

In its behalf we remind the reader of the fundamental facts about finite Abelian groups and their characters. An Abelian group of degree n has a basis $a_1, \ldots, a_t$; these are elements of the group satisfying equations

$$a_1^{n_1} = 1, \ldots, a_t^{n_t} = 1$$

such that every element of the group, and each one only once, is obtained in the form

$$s = a_1^{x_1} \ldots a_t^{x_t} \tag{10.1}$$

if x_i ranges over a complete residue system mod n_i, for instance,

$$x_i = 0, 1, \ldots, n_i - 1 \qquad (i = 1, \ldots, t).$$

The order of the group is

$$n = n_1 \ldots n_t.$$

In other words, a finite Abelian group is the direct product of cyclic groups.

A character of the group is a function $\chi(s)$ which maps the elements s of the group homomorphically upon the unit circle in the complex χ-plane; i.e.,

$$|\chi(s)| = 1, \qquad \chi(ss') = \chi(s) \cdot \chi(s')$$

for any two elements s, s' of the group. It follows at once that

$$\chi(1) = 1, \qquad \chi(s^{-1}) = \chi^{-1}(s) = \overline{\chi}(s).$$

If one puts

$$\chi(a_1) = \zeta_1, \ldots, \chi(a_t) = \zeta_t$$

one must have

$$\zeta_1^{n_1} = 1, \ldots, \zeta_t^{n_t} = 1$$

and for the element (10.1):

$$\chi(s) = \zeta_1^{x_1} \ldots \zeta_t^{x_t}. \tag{10.2}$$

Conversely if ζ_i is any n_ith root of unity ($i = 1, \ldots, t$), then (10.2) defines a character of the group. Hence there are exactly as many distinct group characters,

$$n_1 \ldots n_t = n,$$

as there are elements in the group. A given element s satisfying the equation $\chi(s) = 1$ for all characters χ must be the unit element; one merely has to look at (10.2) to verify this statement.

The character defined by $\chi(s) = 1$ for all group elements s is called the principal character χ_0. Two characters χ_1, χ_2 give rise to a character χ,

$$\chi(s) = \chi_1(s)\chi_2^{-1}(s).$$

Consequently the characters form a multiplicative group in which χ_0 is the unit element. If in (10.2) one takes for one of the ζ's, say ζ_i, a primitive n_ith root of unity and 1 for the other $\zeta_k (k \neq i)$ one realizes at once that the group of characters is isomorphic with the given group.

One more relation is of importance: the sum

$$\sum_s \chi(s) \tag{10.3}$$

extending to all group elements s equals n or 0 according as χ is or is not the principal character. Indeed this sum for the character (10.2) equals

$$\prod_{i=1}^{t} (1 + \zeta_i + \ldots + \zeta_i^{n_i-1}), \tag{10.4}$$

but a n_ith root of unity ζ_i satisfies the equation

$$1 + \zeta_i + \ldots + \zeta_i^{n_i-1} = 0$$

unless $\zeta_i = 1$. Hence at least one of the factors in (10.4) will vanish, except if

$$\zeta_1 = \ldots = \zeta_t = 1, \qquad \chi = \chi_0.$$

By applying (10.3) to the character $\chi_1^{-1}\chi_2$ one obtains the following orthogonality relations:

$$\frac{1}{n} \sum_s \overline{\chi}_1(s)\chi_2(s) = \begin{cases} 1 & \text{for } \chi_1 = \chi_2, \\ 0 & \text{for } \chi_1 \neq \chi_2. \end{cases} \tag{10.5}$$

If one arrays the n characters in definite order in a column $\chi_1, \ldots, \chi_n$ and the values of the argument $s = s_1, \ldots, s_n$ in a row, entering the value of $\chi_i(s_k)$ at the crossing point of the ith row with the kth column, then one obtains a matrix X which according to (10.5) is unitary-orthogonal:

$$\overline{X}X' = n \cdot E$$

(X' transposed of X, E unit matrix).

This relation implies

$$X'\overline{X} = n \cdot E$$

or the following second form of the orthogonality relations:

$$\frac{1}{n} \sum_\chi \chi(s)\overline{\chi}(s') = \begin{cases} 1 & \text{for } s = s', \\ 0 & \text{for } s \neq s'. \end{cases} \tag{10.6}$$

11. Asymptotic Equi-distribution of Ideals Over Their Classes

One would neglect an essential aspect of number fields by overlooking the infinite prime spots; they have a perfectly legitimate place in the arithmetic structure of the fields. But now we go a step further, by introducing calculus, integration, analytic functions. These transcendental methods are powerful although they may with some right be denounced as alien to arithmetic.

Again as in §1 we operate with the lattice $\mathcal{L}$ of the points of integral coördinates in an n-dimensional space. Let $\mathfrak{K}$ be any solid with a Jordan volume V. Let τ be a positive number tending to infinity and consider the lattice $\frac{1}{\tau}\mathcal{L}$, whose points are of the form

$$\left(\frac{a_1}{\tau}, \ldots, \frac{a_n}{\tau}\right) \qquad [a_i \text{ integers}],$$

and its meshes. The Jordan volume is obtained by counting the numbers $N_i(\tau)$ and $N_a(\tau)$ of the meshes which lie entirely in $\mathfrak{K}$ or have a point in common with $\mathfrak{K}$ respectively,

$$N_i(\tau) \leqq N_a(\tau).$$

The volume is defined as the limit

$$\lim_{\tau \to \infty} \left(\frac{1}{\tau}\right)^n N_i(\tau) = \lim_{\tau \to \infty} \left(\frac{1}{\tau}\right)^n N_a(\tau).$$

If one submits the whole configuration to a dilatation $\tau : 1$ one arrives at Gauss' principle: If $\tau\mathfrak{K}$ contains $N(\tau)$ lattice points of $\mathcal{L}\,[N_i(\tau) \leqq N(\tau) \leqq N_a(\tau)]$, then the asymptotic law

$$N(\tau) \sim V \cdot \tau^n$$

holds in the sense that

$$\frac{N(\tau)}{\tau^n} \longrightarrow V \quad \text{with} \quad \tau \to \infty.$$

Still simpler one may write

$$N(\tau) \sim V(\tau\mathfrak{K}). \tag{11.1}$$

If $\mathfrak{K}$ is described by analytic conditions the error in (11.1) will be $O(\tau^{n-1})$, i.e., $\leqq$ Const. τ^{n-1}.

We shall employ Gauss' principle to derive an asymptotic formula for the number $T(C;t)$ of integral ideals in a given class C whose norm $N\mathfrak{a} \leqq t$.

Theorem IV 11, A.

$$T(C;t) \sim \mu \cdot t \quad \text{for } t \to \infty$$

with the constant

$$\mu = \frac{2^{r+1}\pi^{r_2}}{w} \cdot \frac{R}{\sqrt{|d|}}.$$

R designates the regulator.

Proof. We choose an arbitrary ideal $\mathfrak{j}$ in C^{-1}. Then $\mathfrak{j}\mathfrak{a}$ is a principal ideal (α) divisible by $\mathfrak{j}$ and

$$N(\alpha) \leqq N(\mathfrak{j}) \cdot t.$$

Relative to an integral basis $\iota_1, \ldots, \iota_n$ of $\mathfrak{j}$, the numbers α divisible by $\mathfrak{j}$ are of the form

(11.2) $\quad \iota_1 x_1 + \ldots + \iota_n x_n \quad$ (x_i rational integers).

We denote the form (11.2) by ξ and its absolute values for the r_1 real and r_2 complex infinite prime spots $\mathfrak{p}_i$ $(i = 1, \ldots, r + 1)$ by

$$|\xi|_1, \ldots, |\xi|_{r_1}, \quad |\xi|_{r_1+1}, \ldots, |\xi|_{r+1}$$

with the logarithms

$$\zeta_1 = l_1(\xi), \ldots, \zeta_{r+1} = l_{r+1}(\xi).$$

If one multiplies the variables x_i by a positive real factor τ, then $|\xi|_i$ is multiplied by τ or τ^2 according as $\mathfrak{p}_i$ is real or complex. We therefore introduce

$$l_i = \begin{cases} 1 & (\mathfrak{p}_i \text{ real}) \\ 2 & (\mathfrak{p}_i \text{ complex}) \end{cases}$$

and then have

$$l_i(\xi) \to l_i(\xi) + l_i \log \tau$$

under the substitution (dilatation) $x_i \to \tau x_i$.

Let $\varepsilon_1, \ldots, \varepsilon_r$ now be a basis for the (absolute) units. If we set

(11.3) $$l_i(\xi) = l_i(\varepsilon_1) \cdot z_1 + \ldots + l_i(\varepsilon_r) \cdot z_r + l_i \cdot z,$$

$z_1, \ldots, z_r$ are not affected by the dilatation while z changes into $z + \log \tau$. The following numbers,

$$\alpha \cdot \varepsilon_1^{-u_1} \ldots \varepsilon_r^{-u_r}$$

(u_i ranging over all integral exponents), are all associate with the given number α. Among them we can choose exactly one for which

(11.4) $$0 \leqq z_1 < 1, \ldots, 0 \leqq z_r < 1.$$

A factor ζ which is a wth root of unity still remains arbitrary. Hence in every class of associate numbers there are exactly w reduced ones satisfying (11.4). Thus with the abbreviation

(11.5) $$N(\mathfrak{j}) \cdot t = \tau^n,$$

$w \cdot T$ is the number of reduced numbers α divisible by $\mathfrak{j}$ with a norm $\leqq \tau^n$, or the number of lattice points $(x_1, \ldots, x_n)$ for which the inequalities (11.4) hold together with

(11.6) $$N(\xi) = |\xi|_1 \ldots |\xi|_{r+1} \leqq \tau^n.$$

The relations (11.4) define a cone with the origin as vertex of which (11.6) cuts off a bounded portion $\tau\mathfrak{K}$. The notation indicates that $\tau\mathfrak{K}$ arises from $1\mathfrak{K}$ by the dilatation $\tau : 1$. Gauss' principle yields the asymptotic formula

$$wT \sim V(\tau\mathfrak{K}).$$

To compute the volume of $\tau\mathfrak{K}$ we first pass from the variables x_i to the n linear forms

$$y_\lambda; \quad y'_\mu, \ y''_\mu$$

which are the conjugates of (11.2) within Ω. Their determinant is $(N\mathfrak{j})^2 \cdot d$. We cut $\tau\mathfrak{K}$ into 2^{r_1} parts according to the various combinations of signs for the r_1 real linear forms y_λ. Because of symmetry each part contributes the same amount to the total volume, thus accounting for the factor 2^{r_1} in the following formula (11.7). The area element of a complex y-plane,

$$\begin{vmatrix} dy, & d\bar{y} \\ \delta y, & \delta\bar{y} \end{vmatrix},$$

in terms of polar coördinates r, φ is

$$-2i r dr\, \delta\varphi \qquad (\text{with } d\varphi = 0, \quad \delta r = 0),$$

and hence an integrand depending on r alone carries the factor

$$-2\pi i \cdot d(r^2).$$

Therefore

$$(11.7) \quad V(\tau\mathfrak{k}) = \frac{2^{r_1}(2\pi)^{r_2}}{N\mathfrak{j} \cdot \sqrt{|d|}} \cdot \int \cdots \int d|\xi|_1 \cdots d|\xi|_{r+1}.$$

After introducing the logarithms ζ_i of the absolute values $|\xi|_i$ the last integral turns into

$$\cdot \int \cdots \int e^{\zeta_1 + \cdots + \zeta_{r+1}} d\zeta_1 \cdots d\zeta_{r+1}$$

and by the linear substitution (11.3) changes further into

$$(11.8) \quad \text{abs.}\ \left| l_i(\varepsilon_1), \ldots, l_i(\varepsilon_r), l_i \right| \cdot \int \cdots \int e^{nz} dz_1 \cdots dz_r\, dz.$$

The domain of integration is now described by (11.4) and (11.6), or

$$0 \leqq z_1 < 1, \ldots, 0 \leqq z_r < 1; \ -\infty < z \leqq \log \tau.$$

Hence the integration in (11.8) may be carried out and yields

$$\frac{1}{n} \cdot e^{n \log \tau} = \frac{\tau^n}{n}.$$

The determinant in front of (11.8) is computed by adding all preceding rows to the last, the $(r + 1)^{\text{th}}$, row whereby it changes into

$$\begin{vmatrix} l_1(\varepsilon_1), \ldots, & l_1(\varepsilon_r), & l_1 \\ \cdots & \cdots & \cdots \\ l_r(\varepsilon_1), \ldots, & l_r(\varepsilon_r), & l_r \\ 0 \quad , \ldots, & 0 \quad , & n \end{vmatrix} = n \cdot R,$$

R being the regulator of k. Thus

$$V(\tau \mathfrak{k}) = \frac{2^{r+1}\pi^{r_2}}{N\mathfrak{j} \cdot \sqrt{|d|}} \cdot R\tau^n.$$

By reinstating the expression (11.5) for τ^n the factor $N\mathfrak{j}$ drops out and we are left with the asymptotic formula upheld by our theorem, or even with the sharper estimate

$$T(C;t) = \mu t + O\left(t^{1-\frac{1}{n}}\right).$$

It is not the value of the constant μ that matters greatly, but these two facts:

(1) the number of integral ideals of norm $\leqq t$ in a class is of order t, just as the number of positive integers $\leqq t$ in the rational field is of order t;

(2) μ is independent of the class under consideration, which amounts to asymptotic equi-distribution of the ideals over their H classes.

Summing over all classes we find:

Theorem IV 11, B. *The number $T(t)$ of integral ideals with a norm $\leqq t$ is asymptotically*

$$(11.9) \qquad \sim \mu H \cdot t,$$

the error being $O(t^{1-\frac{1}{n}})$.

12. ζ-function and Related Dirichlet Series

One knows that the series

$$\zeta(s) = \sum_{\nu=1}^{\infty} \frac{1}{\nu^s}$$

extending over the natural numbers $\nu = 1, 2, \ldots$ converges for any real value $s > 1$ and converges uniformly for $s \geqq \sigma_0 > 1$. Since for complex s,

$$\left|\frac{1}{\nu^s}\right| = \frac{1}{\nu^\sigma} \qquad (\sigma = \mathcal{R}s),$$

the series converges in the half-plane $\mathcal{R}s > 1$ and uniformly for $\mathcal{R}s \geqq \sigma_0 > 1$ and thus represents an analytic function of s in the half-plane $\mathcal{R}s > 1$. Euler first observed that it may be written as an infinite product involving the prime numbers p, namely

$$\zeta(s) = \prod_p \frac{1}{1 - p^{-s}} \tag{12.1}$$

Indeed after expanding

$$\frac{1}{1 - p^{-s}} = 1 + p^{-s} + p^{-2s} + p^{-3s} + \ldots$$

this relation is merely a condensed analytic form of the fundamental arithmetical theorem that every integer ν is uniquely representable as a power product of distinct primes. To make the proof complete, first assume s to be real (and >1). The product (12.1) extending over the first h prime numbers $p_1, \ldots, p_h$ equals

$$\Sigma \frac{1}{\nu^s}$$

with ν running over those integers which are power products of $p_1, \ldots, p_h$. The sum contains all the terms $\nu = 1, \ldots, h$ and hence differs from $\zeta(s)$ by less than the remainder

$$\sum_{\nu=h+1}^{\infty} \frac{1}{\nu^s}$$

thus tending to $\zeta(s)$ with $h \longrightarrow \infty$. The same is true for complex s:

$$\left|\zeta(s) - \prod_{i=1}^{h} \frac{1}{1 - p_i^{-s}}\right| \leqq \sum_{\nu=h+1}^{\infty} \frac{1}{\nu^\sigma}$$

The product shows that $\zeta(s) \neq 0$ in the whole half-plane $\mathcal{R}s > 1$. Taking the logarithm one gets this formula

$$\log \zeta(s) = \sum_p \log \frac{1}{1 - p^{-s}} = \sum_p (p^{-s} + \frac{1}{2} p^{-2s} + \frac{1}{3} p^{-3s} + \ldots)$$

from which Euler drew the conclusion that the sum

$$\sum_p \frac{1}{p} \tag{12.2}$$

over the prime numbers p diverges, thus incidentally confirming what Éuclid had already known, that there are infinitely many prime numbers. The divergence of (12.2) reveals something about the density with which the primes are distributed among all numbers; for instance they are not spread as thinly as the square numbers.

Dirichlet derived some surprising arithmetical results, particularly about the existence and distribution of prime numbers, from a study of this kind of series,

$$\sum_{\nu=1}^{\infty} \frac{a_\nu}{\nu^s} \tag{12.3}$$

which are therefore called Dirichlet series. Riemann discovered the fact that $\zeta(s)$ is a meromorphic function in the entire s-plane and he developed a functional equation connecting $\zeta(1 - s)$ with $\zeta(s)$. Moreover he indicated how to use the analytic behaviour of $\zeta(s)$ for a thorough study of the distribution of prime numbers. More than half a century later his paper became the starting point for the modern theory of entire functions (Hadamard) by which some of Riemann's conjectures were proved while others still defy the skill of the analyst.

Here we consider the ζ-function $\zeta(s) = \zeta_k(s)$ of a field k over $\mathfrak{g}$ which is defined as the sum

$$\zeta_k(s) = \sum_{\mathfrak{a}} \frac{1}{(N\mathfrak{a})^s}$$

extending over all integral ideals $\mathfrak{a}$ of k. If a_ν denotes the number of ideals $\mathfrak{a}$ of norm ν, then $\zeta_k(s)$ is the Dirichlet series (12.3). By partial summation one gets

$$\sum_{\nu=1}^{N} \frac{a_\nu}{\nu^s} = a_1\left(\frac{1}{1^s} - \frac{1}{2^s}\right) + (a_1 + a_2)\left(\frac{1}{2^s} - \frac{1}{3^s}\right) + \dots$$

$$+ (a_1 + \dots + a_{N-1})\left(\frac{1}{(N-1)^s} - \frac{1}{N^s}\right) + R_N,$$

$$R_N = (a_1 + \dots + a_N) \cdot \frac{1}{N^s}.$$

Since

$$a_1 + \ldots + a_t = T(t) \sim \mu H \cdot t,$$

$R_N \to 0$ with $N \to \infty$ for $\mathcal{R}s > 1$, and thus

$$\zeta_k(s) = \sum_{\nu=1}^{\infty} T(\nu)\left\{\frac{1}{\nu^s} - \frac{1}{(\nu+1)^s}\right\}. \tag{12.4}$$

The derivative of x^{-s} is $-s \cdot x^{-s-1}$, hence

$$\frac{1}{\nu^s} - \frac{1}{(\nu+1)^s} = s\int_{\nu}^{\nu+1} \frac{dx}{x^{s+1}}.$$

The infinite sum (12.4) is therefore replaceable by the integral

$$s \cdot \int_1^{\infty} \frac{T(t)dt}{t^{s+1}}.$$

The asymptotic law (11.9) implies

$$\zeta_k(s) \sim s \cdot \int_1^{\infty} \frac{\mu Ht \cdot dt}{t^{s+1}} = \mu H \cdot \frac{s}{s-1} \sim \frac{\mu H}{s-1}$$

in the sense that for real $s > 1$ tending to 1,

$$(s-1) \cdot \zeta_k(s) \to \mu H. \tag{12.5}$$

Putting $T^*(t) = T(t) - \mu Ht$, one has simply to prove that

$$T^*(t)/t \to 0 \quad \text{for} \quad t \to \infty$$

entails

$$(s-1)\int_1^{\infty} \frac{T^*(t)}{t^{s+1}}\, dt \to 0 \quad \text{for} \quad s \to 1. \tag{12.6}$$

For an arbitrarily given $\varepsilon > 0$ one splits

$$\int_1^{\infty} \quad \text{into} \quad \int_1^u + \int_u^{\infty}$$

such that

$$\left|\frac{T^*(t)}{t}\right| \leq \varepsilon \quad \text{for} \quad t \geq u.$$

Then

$$\left| \int_u^\infty \frac{T^*(t)}{t^{s+1}}\, dt \right| \leqq \varepsilon \int_u^\infty \frac{dt}{t^s} \leqq \varepsilon \int_1^\infty \frac{dt}{t^s} = \frac{\varepsilon}{s-1} .$$

After having fixed u, the other part

$$(s-1) \cdot \int_1^u \frac{T^*(t)}{t^{s+1}}\, dt$$

will be less than ε for s sufficiently near to 1, and then

$$(s-1) \int_1^\infty \frac{T^*(t)}{t^{s+1}}\, dt < 2\varepsilon .$$

The same relation (12.6) holds good when s approaches 1 in the complex s-plane from within a fixed angle $<\pi$ with its vertex in 1 and the real axis as its center line. In making use of the sharper estimate

$$f(t) = \mu H \cdot t + O\left(t^{1-\frac{1}{n}}\right)$$

one finds that $\zeta_k(s)$ is a regular analytic function in the half-plane

$$\mathcal{R}s > 1 - \frac{1}{n}$$

except for a pole of order 1 at the point s = 1 with the residue μH.

(12.5) is merely a weakened restatement of Theorem IV 11, B. The weakening would be foolish were it not for the connection of the ζ-function with the prime ideals $\mathfrak{p}$ as established by Euler's formula

$$\zeta_k(s) = \prod_{\mathfrak{p}} \frac{1}{1-(N\mathfrak{p})^{-s}} .$$

This formula which is proved exactly as for the rational field gives rise to the following Dirichlet series for the logarithm of the ζ-function,

$$\log \zeta_k(s) = \sum_{\mathfrak{p}} \left\{ (N\mathfrak{p})^{-s} + \frac{1}{2}(N\mathfrak{p})^{-2s} + \frac{1}{3}(N\mathfrak{p})^{-3s} + \dots \right\} .$$

As usually we set

$$N\mathfrak{p} = P = p^f (\geqq p).$$

For real $s > \frac{1}{2}$ we find

$$\frac{1}{2} p^{-2s} + \frac{1}{3} p^{-3s} + \ldots < \frac{1}{2} (p^{-2s} + p^{-3s} + \ldots)$$

$$= \frac{1}{2} \frac{p^{-2s}}{1 - p^{-s}} < 2 \cdot p^{-2s},$$

and since there are at most n prime ideals $\mathfrak{p}$ going into p,

$$g(s) = \Sigma \left\{ \frac{1}{2} (N\mathfrak{p})^{-2s} + \frac{1}{3} (N\mathfrak{p})^{-3s} + \ldots \right\}$$

$$\leqq 2n \cdot \sum_p p^{-2s} \leqq 2n \cdot \sum_{\nu=1}^{\infty} \nu^{-2s}.$$

Hence $\log \zeta_k(s)$ differs from $\sum_{\mathfrak{p}} (N\mathfrak{p})^{-s}$ by a function g(s) which stays bounded in the neighborhood of s = 1, and therefore:

Theorem IV 12, A.

(12.7) $$\sum_{\mathfrak{p}} (N\mathfrak{p})^{-s} \sim \log \frac{1}{s - 1}$$

in the sense that the difference of left and right members stays bounded or even tends to a finite limit $\log(\mu H)$ if $s > 1$ tends to 1.

The proof shows at once that the relation remains valid if the summation is limited to prime ideals $\mathfrak{p}$ of degree 1.

The relation (12.7) for the rational ground field,

(12.7') $$\sum_p p^{-s} \sim \log \frac{1}{s - 1}$$

is the quintessence of Euler's discovery. Could one, for $\log \zeta(s)$, retrace the steps which for the ζ-function itself led from (11.9) to (12.5) one would infer from (12.7) this asymptotic law for the number $\pi(t)$ of prime numbers $\leqq t$:

(12.8) $$\pi(t) \sim \frac{t}{\log t}.$$

Indeed

$$\int_2^{\infty} \frac{dx}{x^s \log x} \sim \log \frac{1}{s - 1}$$

since its derivative with respect to s equals

$$-\int_2^{\infty} \frac{dx}{x^s} = -\frac{2^{1-s}}{s-1} \sim -\frac{1}{s-1}.$$

This transition from (12.7) to (12.8) is the main subject of the so-called analytic theory of numbers. Here we will be content with the relations (12.7) and (12.7'); this complaisant attitude induces us to ascribe a Dirichlet density δ to any set of prime ideals $\mathfrak{p}$ if

$$\Sigma(N\mathfrak{p})^{-s} \sim \delta \cdot \log \frac{1}{s-1},$$

the sum extending over the prime ideals of the set under consideration. For instance the prime ideals of degree 1 have the Dirichlet density 1, while the prime ideals of higher degrees have the Dirichlet density 0.

In passing we mention that Hecke has proved that $\zeta_k(s)$, just as Riemann's ζ-function, is meromorphic in the whole s-plane and satisfies a similar functional equation.

Up to now we have only used a part of our results concerning the density of ideals, namely we have summed over all classes ignoring the fact of equi-distribution. We make up for this negligence by forming the ζ-function of a class C,

$$\zeta(C;s) = \sum_{\mathfrak{a} \text{ in } C} \frac{1}{(N\mathfrak{a})^s}$$

and then get the asymptotic law

$$\zeta(C;s) \sim \frac{\mu}{s-1} \quad \text{for} \quad s \longrightarrow 1. \tag{12.9}$$

The transition to prime ideals becomes possible only after introducing characters. If $\chi(C)$ is a character of the classes we put $\chi(\mathfrak{a}) = \chi(C)$ for every ideal $\mathfrak{a}$ in C, so that

(12.10)

$$\chi(\mathfrak{a}\mathfrak{b}) = \chi(\mathfrak{a}) \cdot \chi(\mathfrak{b}) \text{ and } \chi(\mathfrak{a}) = \chi(\mathfrak{a}^*) \text{ if } \mathfrak{a} \simeq \mathfrak{a}^*.$$

We now form the so-called L-series each of which corresponds to a definite character χ:

$$L(\chi;s) = \sum_C \chi(C) \cdot \zeta(C;s) = \Sigma \frac{\chi(\mathfrak{a})}{(N\mathfrak{a})^s}. \tag{12.11}$$

The last sum again extends to all integral ideals. As one sees they are simple linear combinations of the $\zeta(C;s)$. Conversely one easily passes back from the $L(\chi;s)$ to the $\zeta(C;s)$: owing to the orthogonality relations (10.6), the equations (12.11) allow the solution

$$\zeta(C;s) = \frac{1}{H}\sum_{\chi}\overline{\chi}(C) \cdot L(\chi;s)$$

with the sum extending over all H characters χ. The asymptotic laws (12.9) are now replaced by

$$L(\chi;s) \sim \begin{cases} \dfrac{\mu H}{s-1} & \text{if } \chi \text{ is the principal character } \chi_0, \\ \dfrac{0}{s-1} & \text{in all other cases.} \end{cases}$$

Hence unless χ is principal, $L(\chi;s)$ is a regular analytic function in the half-plane $\mathcal{R}s > 1 - \frac{1}{n}$, even at $s = 1$. The value of this analytic function at the point $s = 1$ will be denoted by $L(\chi;1)$.

Because of (12.10), $L(\chi;s)$ may be written as an Euler product over all prime ideals $\mathfrak{p}$,

$$L(\chi;s) = \prod_{\mathfrak{p}} \frac{1}{1 - \chi(\mathfrak{p})(N\mathfrak{p})^{-s}}.$$

We obtain the following relations which refer to a real $s > 1$ approaching 1:

$$(12.12)\quad \sum_{\mathfrak{p}} \frac{\chi(\mathfrak{p})}{(N\mathfrak{p})^s} \begin{cases} \sim \log \dfrac{1}{s-1} & \text{for } \chi = \chi_0, \\ < \text{Const. for all other characters,} \end{cases}$$

Const. meaning a number independent of s. Notice the one-sided boundedness! In order to make it two-sided,

$$\left| \sum_{\mathfrak{p}} \frac{\chi(\mathfrak{p})}{(N\mathfrak{p})^s} \right| \leq \text{Const.} \qquad (\chi \neq \chi_0)$$

one would have to know that $L(\chi;1)$ is not only finite but also $\neq 0$. Taking this for granted we can, in the manner described before, go back from

$$\sum \frac{\chi(\mathfrak{p})}{(N\mathfrak{p})^s} = \sum_{C}\left\{\chi(C) \cdot \sum_{\mathfrak{p} \text{ in } C} \frac{1}{(N\mathfrak{p})^s}\right\}$$

to the individual terms in the right side sum over C and get

$$\sum_{\mathfrak{p} \text{ in } C} \frac{1}{(N\mathfrak{p})^s} \sim \frac{1}{H} \log \frac{1}{s-1}$$

Taking the term density in its Dirichlet sense we may state this result briefly as follows:

Theorem IV 12, B. *The prime ideals are equidistributed over the H classes of ideals.*

In particular we may assert:

Corollary. *In every class there are (infinitely many) prime ideals.*

One sees how the Dirichlet series and the characters enable one to pass from the ideals to the prime ideals.

The gap in our proof can be filled out in two essentially different ways: One infers the two-sided from the one-sided boundedness of (12.12) either by a closer analytic study of Dirichlet series, or, and this is Dirichlet's own ingenious method, one ties up the values $L(\chi;1)$ with a class number, which by its very nature is positive. We illustrate Dirichlet's method by proving his famous theorem that for a given modulus (rational integer) m and a preassigned residue a prime to m there are infinitely many prime numbers $p \equiv a$ (m).

13. Prime Numbers in Residue Classes Modulo m

The residues a modulo m which are prime to m form an Abelian group of degree $\varphi(m)$. Let $\chi(a)$ designate any of its characters. For a given a the correspondence $\chi \longrightarrow \chi(a)$ defines a homomorphic mapping of the group of characters upon the unit circle, or more precisely, upon a subgroup of the group of $\varphi(m)^{\text{th}}$ roots of unity. This subgroup necessarily consists of all f^{th} roots of unity where f is a certain divisor of $\varphi(m)$, and each of these roots η is taken-on g times,

$$f \cdot g = \varphi(m).$$

Hence the auxiliary formula

$$(13.1) \qquad \prod_{\chi}\left\{1 - \chi(a)z\right\} = \prod_{\eta}(1 - \eta z)^g = (1 - z^f)^g.$$

f is the least exponent such that $(\chi(a))^f = 1$ for every character χ, or $\chi(a^f) = 1$, or the least exponent for which $a^f \equiv 1\ (m)$.

This formula serves to compute the ζ-function $\zeta_K(s)$ of the m-cyclotomic field K over $\mathfrak{g}$ which we have studied in Ch. III, §12,

$$\zeta_K(s) = \prod_{\mathfrak{p}} \frac{1}{1 - (N\mathfrak{p})^{-s}} \cdot \tag{13.2}$$

Any rational prime number p not dividing m splits in K into g distinct prime ideals $\mathfrak{p}_1, \ldots, \mathfrak{p}_g$ of degree f where f is the least exponent such that $p^f \equiv 1\ (m)$ [Theorem III 12, E]. Hence these g prime ideals contribute to the product (13.2) the factor

$$\frac{1}{(1 - p^{-fs})^g}$$

which, according to (13.1) with $z = p^{-s}$, equals

$$\prod_{\chi} \frac{1}{1 - \chi(p) \cdot p^{-s}}$$

The product'

$$\prod_{p} \frac{1}{1 - \chi(p) \cdot p^{-s}}$$

extending over all prime numbers p not dividing m equals

$$L(\chi; s) = \sum_{\nu} \frac{\chi(\nu)}{\nu^s}$$

with ν ranging over all positive integers prime to m. The t primes $l_1, \ldots, l_t$ going into m are of little influence. If we take their contribution into account by the same Theorem III 12, E, and with the notations used there, we obtain the following formula

$$\zeta_K(s) = \prod_{j=1}^{t} \frac{1}{(1 - l_j^{-f_j s})^{g_j}} \cdot \prod_{\chi} L(\chi; s).$$

On the other hand the ζ-function of the rational field $\mathfrak{g}$ equals

$$\zeta(s) = \prod_{j=1}^{t} \frac{1}{1 - l_j^{-s}} \cdot \prod_{p} \frac{1}{1 - p^{-s}} \quad \text{(p prime to m)}$$

$$= \prod_{j=1}^{t} \frac{1}{1 - l_j^{-s}} \cdot L(\chi_0;s),$$

therefore the quotient

$$\text{(13.3)} \qquad \frac{\zeta_K(s)}{\zeta(s)} = V(s) \cdot \prod_{\chi \neq \chi_0} L(\chi;s).$$

The elementary function

$$V(s) = \prod_{j=1}^{t} \frac{1 - l_j^{-s}}{(1 - l_j^{-f_j s})^{g_j}}$$

is regular and $\neq 0$ for $\mathcal{R}\, s > 0$. Incidentally $V(s) = 1$ if m is a prime power.

For any character $\chi \neq \chi_0$ the sum

$$\sum_a \chi(a) = 0$$

when being extended over all $\varphi(m)$ distinct residues a prime to m. The Dirichlet series

$$\text{(13.4)} \quad L(\chi;s) = \sum \frac{\chi(\nu)}{\nu^s} = \sum_{\nu=1}^{\infty} \frac{b_\nu}{\nu^s} \quad [b_\nu = 0 \text{ unless } (\nu,m) = 1]$$

has periodic coefficients with the period m, and by the last remark the sum of its ν first coefficients

$$b_1 + b_2 + \ldots + b_\nu$$

is a bounded function of ν. Partial summation therefore ensures the convergence of (13.4) in the entire half-plane $\mathcal{R}\, s > 0$.

From the asymptotic laws

$$\zeta_K(s) \sim \frac{\mu H}{s - 1}, \qquad \zeta(s) \sim \frac{1}{s - 1}$$

we thus infer, by passing to the limit $s = 1$, the highly interesting formula

$$\text{(13.5)} \qquad \mu H = V(1) \cdot \prod_{\chi \neq \chi_0} L(\chi;1).$$

It proves at once

$$L(\chi;1) \neq 0 \quad \text{for} \quad \chi \neq \chi_0$$

and thus results in Dirichlet's

Theorem IV 13, A. *There are infinitely many prime numbers p which are congruent a modulo m if a is prime to m; more precisely, the prime numbers are equi-distributed over the $\varphi(m)$ residue classes:*

$$\sum_{p \equiv a\ (m)} p^{-s} \sim \frac{1}{\varphi(m)} \log \frac{1}{s-1}$$

This is one direction toward which one can turn the formula (13.5). Another perhaps even more important aspect is that it provides an analytic tool for computing the class number H of the cyclotomic field. We will carry this out for the quadratic rather than the cyclotomic field.

Adaptation of this method to the situation encountered in the previous section would require the construction of the class field over k, an Abelian field of degree H whose Galois group is isomorphic to the class group of k and which stands in the same relationship to the classes of ideals in k as the m-cyclotomic field stands to the residue classes modulo m of numbers in $\mathfrak{Q}$.

14. ζ-function of Quadratic Fields, and Their Application

Let d be the discriminant of the quadratic field $\varkappa = \mathfrak{Q}(\sqrt{d})$. We know the meaning of the quadratic residue symbol $\left(\frac{d}{p}\right)$ for any odd prime number p not dividing d. For the sake of convenience, we put

$$\left(\frac{d}{p}\right) = 0 \quad \text{if } d : p,$$

and if 2 does not go into d and hence $d \equiv 1\ (4)$:

$$\left(\frac{d}{2}\right) = (-1)^{\frac{d^2-1}{8}} = \begin{cases} +1 & \text{for } d \equiv \pm 1\ (8), \\ -1 & \text{for } d \equiv \pm 5\ (8). \end{cases}$$

The contribution of the prime number p to the product

$$\zeta_2(s) = \Pi \frac{1}{1 - (N\mathfrak{p})^{-s}}$$

is

$$\frac{1}{1 - p^{-s}} \quad \text{if } (p) = \mathfrak{p}^2 \qquad \text{or } d : p,$$

$$\left(\frac{1}{1 - p^{-s}}\right)^2 \quad \text{if } (p) = \mathfrak{p}_1 \mathfrak{p}_2 \quad \text{or } \left(\frac{d}{p}\right) = 1,$$

$$\frac{1}{1 - p^{-2s}} \quad \text{if } (p) = \mathfrak{p} \qquad \text{or } \left(\frac{d}{p}\right) = -1,$$

hence in every case

$$\frac{1}{1 - p^{-s}} \cdot \frac{1}{1 - \left(\frac{d}{p}\right)p^{-s}}.$$

We find

$$\zeta_2(s) = \prod_p \left\{\frac{1}{1 - p^{-s}} \cdot \frac{1}{1 - \left(\frac{d}{p}\right)p^{-s}}\right\} = \zeta(s) \cdot \prod_p \frac{1}{1 - \left(\frac{d}{p}\right)p^{-s}}$$

consequently

$$(14.1) \qquad \prod_p \frac{1}{1 - \left(\frac{d}{p}\right)p^{-s}} \longrightarrow \mu H \quad \text{for } s \longrightarrow 1.$$

We infer from (14.1) in a familiar way that

$$\sum_s \left(\frac{d}{p}\right)p^{-s} \quad \text{stays finite for } s \longrightarrow 1,$$

or in other words:

If a is a rational integer which is not a perfect square then the sum

$$(14.2) \qquad \Sigma' \left(\frac{a}{p}\right)p^{-s}$$

extending over all odd prime numbers p not dividing a stays finite for $s \longrightarrow 1$.

An immediate consequence is the existence of infinitely many prime numbers p for which $\left(\frac{a}{p}\right) = -1$. But we can get much farther in this direction. Let $\delta_1, \ldots, \delta_t$ be given signs ± 1 and $a_1, \ldots, a_t$ given integers such that

$a_1^{u_1} \ldots a_t^{u_t}$ is a perfect square only if all the exponents u are even.

Theorem IV 14, A. *Under the circumstances just described there are infinitely many odd prime numbers p not dividing* $a_1, \ldots, a_t$ *for which the simultaneous equations*

$$\left(\frac{a_1}{p}\right) = \delta_1, \ldots, \left(\frac{a_t}{p}\right) = \delta_t \tag{14.3}$$

hold. In fact their Dirichlet density is $1/2^t$.

Proof. Consider the sum

$$\sum_p{}' \left\{1 + \delta_1^{-1}\left(\frac{a_1}{p}\right)\right\} \cdots \left\{1 + \delta_t^{-1}\left(\frac{a_t}{p}\right)\right\} \cdot \frac{1}{p^s}$$

extending to all odd prime numbers p not dividing $a_1, \ldots, a_t$. In carrying out multiplication of the t factors $\{\ \}$ and resorting to (14.2) one finds this sum to be

$$\sim \log \frac{1}{s-1} \cdot$$

On the other hand the product of the t factors $\{\ \}$ is 0 unless all the equations (14.3) hold in which case its value is 2^t.

This is in line with the investigation of the previous section. But now we turn the other way and derive from (14.1) an explicit finite expression for the class number H, We study the special case

$$d = \pm l \equiv 1 \pmod 4)$$

where d contains only one prime number l. Our quadratic field then is a subfield of the l-cyclotomic field, and in this way we previously proved the reciprocity law

$$\left(\frac{\pm l}{p}\right) = \left(\frac{p}{l}\right)$$

which is even true for p = 2. Hence

$$\prod_p \frac{1}{1 - \left(\frac{d}{p}\right)p^{-s}} = \prod_{p \neq l} \frac{1}{1 - \left(\frac{p}{l}\right)p^{-s}} = \Sigma\left(\frac{n}{l}\right)n^{-s} \quad [n > 0,\ (n,l) = 1]$$

The coefficients of this Dirichlet series are periodic,

$$\left(\frac{n}{l}\right) = \left(\frac{n'}{l}\right) \quad \text{if} \quad n \equiv n' \pmod{l}$$

and since there are as many quadratic residues mod l as non-residues their sum

$$\sum_n \left(\frac{n}{l}\right)$$

extending over a full period vanishes. Consequently the Dirichlet series converges and represents a regular analytic function for $\mathcal{R}\, s > 0$. Our task is to compute its value for $s = 1$.

To that purpose we make use of the Γ-function whose definition at once yields the equation

$$\Gamma(s) \cdot n^{-s} = \int_0^\infty e^{-nt} \cdot t^{s-1}dt.$$

Setting $e^{-t} = x$ we have to compute (for $0 < x < 1$)

$$\Sigma\left(\frac{n}{l}\right)x^n = \sum_{n=1}^{l-1} \left(\frac{n}{l}\right)x^n \cdot \frac{1}{1 - x^l}.$$
$$[n > 0,\ (n,l) = 1]$$

We put

$$f(x) = \sum_{n=1}^{l-1}\left(\frac{n}{l}\right)x^n.$$

This polynomial of degree $l - 1$ contains the factor x, and as it vanishes for $x = 1$, also the factor $x - 1$. Writing

$$dt = -dx/x$$

we obtain

$$\mu H = \int_0^1 \frac{f(x)}{x} \cdot \frac{dx}{1 - x^l}$$

According to well-known recipes one carries out the integration by decomposing the integrand into partial fractions.

If ζ runs over the lth roots of unity one thus gets

$$\frac{f(x)}{x} \cdot \frac{1}{x^l - 1} = \sum_\zeta \frac{f(\zeta)}{\zeta} \cdot \frac{1}{l\zeta^{l-1}} \cdot \frac{1}{x - \zeta} = \frac{1}{l} \sum_\zeta \frac{f(\zeta)}{x - \zeta} .$$

The root $\zeta = 1$ may be omitted. For

$$\zeta = e^{2\pi i a/l} = e\left(\frac{a}{l}\right) \qquad [1 \leqq a \leqq l - 1]$$

one has

$$f(\zeta) = \sum_{n=1}^{l-1} \left(\frac{n}{l}\right) e\left(\frac{an}{l}\right) = \left(\frac{a}{l}\right) \cdot \sum_{n=1}^{l-1} \left(\frac{an}{l}\right) e\left(\frac{an}{l}\right) = \left(\frac{a}{l}\right) \sum_n \left(\frac{n}{l}\right) e\left(\frac{n}{l}\right) .$$

The latter sum was found in (III 11, 8) to have the value $\sqrt{\pm l}$ (where the sign of the square root remains doubtful). Hence

$$\mu H = -\frac{\sqrt{\pm l}}{l} \cdot \sum_{a=1}^{l-1} \left(\frac{a}{l}\right) \int_0^1 \frac{dx}{x - e(a/l)} . \tag{14.4}$$

The integral in the last sum is

$$\log \frac{1 - \zeta}{-\zeta} = \log (1 - \zeta^{-1}). \tag{14.5}$$

If from now on ζ denotes the primitive l^{th} root of unity $e(1/l)$, the sum in (14.4) is the logarithm of

$$\eta = \prod_a (1 - \zeta^{-a}) / \prod_b (1 - \zeta^{-b})$$

where a runs over all quadratic residues and b over all non-residues. It is evident that η is a unit in the cyclotomic field, and as it is invariant with respect to the substitution $\zeta \longrightarrow \zeta^x$ (x any quadratic residue) it lies in our quadratic subfield. We call η the cyclotomic unit of $\varkappa$. With the integration in (14.4) extending along the real segment $0 \leqq x \leqq 1$ the precise value of our logarithm (14.5) is easily determined; apart from a summand not depending on a, we find

$$\log \left(\sin \frac{\pi a}{l}\right) - i \cdot \frac{\pi a}{l} \qquad (i = \sqrt{-1})$$

where the first term indicates the real logarithm of the positive number $\sin \frac{\pi a}{l}$.

In our further computations we separate real and imaginary quadratic fields:

$$\text{(real) } l \equiv 1 \pmod 4, \quad d = l;$$

$$\text{(imaginary) } l \equiv -1 \pmod 4, \quad d = -l.$$

Real Case. In this case our sum

$$\sum_{n=1}^{l-1} \left(\frac{n}{l}\right) \left(\log \sin \frac{\pi n}{l} - i \cdot \frac{\pi n}{l}\right) \tag{14.6}$$

must be real, or

$$\sum_{n=1}^{l-1} \left(\frac{n}{l}\right) \cdot n = 0.$$

Indeed because of

$$\left(\frac{-1}{l}\right) = 1$$

the substitution $n \longrightarrow l - n$ carries

$$\sum_{n=1}^{l-1} \left(\frac{n}{l}\right) \cdot n \quad \text{into} \quad \sum_{n=1}^{l-1} \left(\frac{n}{l}\right) \cdot (l - n), \quad \text{or}$$

$$2 \sum_{n=1}^{l-1} \left(\frac{n}{l}\right) \cdot n = l \cdot \sum_{n=1}^{l-1} \left(\frac{n}{l}\right) = 0.$$

η is real and positive. There exists a basic unit ε such that all units of our real quadratic field $\varkappa$ are of the form $\pm\varepsilon^h$. We may assume ε to be positive and >1. Then we find

$$\mu = \frac{2 \log \varepsilon}{\sqrt{l}} \quad \text{and} \quad H = \frac{\log \eta}{2 \log \varepsilon}$$

or

$$\eta = \varepsilon^{2H}.$$

As a unit in $\varkappa$, η must be a power of ε; we learn that the exponent is even and twice the class number. The result is of surprising beauty and simplicity. (The sign remains doubtful, the question whether $\eta > 1$ or <1 must here be left open.)

Imaginary Case. In this case our sum (14.6) must be pure-imaginary, and indeed

$$\prod_a \sin \frac{\pi a}{l} \Big/ \prod_b \sin \frac{\pi b}{l} = 1$$

[a quadr. res., b non-res.]

since $b = l - a$.

$$\mu = \frac{2\pi}{w} \cdot \frac{1}{\sqrt{l}} = \frac{\pi}{\sqrt{l}} \text{ (except for } l = 3 \text{ where } w = 6).$$

The result is

$$H = \frac{\Sigma b - \Sigma a}{l}$$

where a runs over all quadratic residues, b over all non-residues in the interval $0 < x < l$. The sign remains doubtful. For $l = 3$ (field of the third root of unity) one finds the trivial result $H = 1$.

By a further simplification one can get rid of the denominator l. We distinguish two cases:

$$l \equiv 7 \quad \text{or} \quad \equiv 3 \pmod 8.$$

(Case a) $l \equiv 7 \pmod 8$. Because of

$$\left(\frac{2}{l}\right) = 1,$$

2a runs over all quadratic residues if a does; and 2a or $2a - l$ lies in the interval $0 < x < l$ according as $a < \frac{1}{2}l$ or $a > \frac{1}{2}l$. Similarly for b. Thus we get

$$\Sigma a = 2 \cdot \Sigma a - l \cdot \underset{(a > \frac{1}{2}l)}{\Sigma 1},$$

$$\Sigma b = 2 \cdot \Sigma b - l \cdot \underset{(b > \frac{1}{2}l)}{\Sigma 1},$$

$\Sigma b - \Sigma a = l$ times excess of number of quadratic non-residues over residues in the interval from $\frac{1}{2}l$ to l, therefore

H = excess of quadratic non-residues over residues, or since $-b$ is quadratic residue if b is non-residue,

$$H = \text{excess (pos.-neg. quadr. res.)}$$

with the proviso that all residues mod l are now taken in the interval $-\frac{1}{2} l < x < \frac{1}{2} l$.

(Case b) $l \equiv 3 \pmod 8$. Since now

$$\left(\frac{2}{l}\right) = -1,$$

the sequence 2b yields the quadratic residues, and one finds

$$\Sigma a = 2 \cdot \Sigma b - l \cdot \sum_{b > \frac{1}{2} l} 1,$$

$$\Sigma b = 2 \cdot \Sigma a - l \cdot \sum_{a > \frac{1}{2} l} 1,$$

and thus readily

$$H = \frac{1}{3} \cdot \text{excess (pos.-neg. quadr. res.)}.$$

A non-transcendental derivation of these wondrous results is unknown. Incidentally Gauss succeeded in evaluating the doubtful sign and found it in agreement with the formulas as given here. Hence modulo a prime number $l \equiv 3$ (4) there are more positive than negative quadratic residues between $-\frac{1}{2}l$ and $\frac{1}{2}l$.

We conclude with a simple example, the quadratic field $\mathfrak{K}(\sqrt{-31})$. The number 31 is $\equiv 3 \pmod 4$ and more particularly $\equiv 7$ (8). As residues modulo 31 we use the numbers in the interval $-15 \leqq x \leqq 15$. We wish to determine all quadratic residues in that interval. In making use of the fact that the difference series of the square numbers in their natural order consists of the odd numbers 1, 3, 5,... we readily obtain this sequence

1, 4, 9, -15, -6, 5, -13, 2, -12, 7, -3, -11, 14, 10, 8.

There are thus three more positive than negative quadratic residues between -15 and +15, and hence according to our formula the class number must be 3.

Let us confirm this by direct calculation. In every class there must be an (integral) ideal $\mathfrak{a}$ such that

$$N\mathfrak{a} \leqq \frac{2}{\pi}\sqrt{31} \quad \text{or} \quad N\mathfrak{a} \leqq 3.$$

(2) decomposes into the two conjugate prime ideals

$$\mathfrak{p} = \left(2, \frac{1 + \sqrt{-31}}{2}\right), \qquad \mathfrak{p}' = \left(2, \frac{1 - \sqrt{-31}}{2}\right)$$

Because of $-31 \equiv -1$ (3), -31 is quadratic non-residue modulo 3 and therefore (3) itself is prime ideal. The only ideals whose norms ≤ 3 are therefore

$$1, \quad \mathfrak{p}, \mathfrak{p}'$$

and we can not have more than 3 classes. But the three ideals 1, $\mathfrak{p}$, $\mathfrak{p}'$ are actually inequivalent. Were $\mathfrak{p}$ principal = (γ) the norm of γ would be 2. All norms of integers are of the form

$$\frac{1}{4}\,(x^2 + 31y^2) \tag{14.7}$$

(x and y odd, or x and y even).

Hence 2 is not a norm. Were $\mathfrak{p}' \sim \mathfrak{p}$ one would change

$$(2) = \mathfrak{p}\mathfrak{p}'$$

into a principal ideal by substituting the equivalent $\mathfrak{p}$ for $\mathfrak{p}'$, in other words $\mathfrak{p}^2$ would be principal. This number would have the norm 4, but as (14.7) shows the only integers of norm 4 are ± 2, hence

$$(2) = \mathfrak{p}^2, \qquad \mathfrak{p} = \mathfrak{p}'$$

which is glaring nonsense.

15. Norm Residues in Quadratic Fields

I devote the last two sections to a concept which has proved fundamental for the theory of class fields, namely that of norm residue and by its means proceed to the formulation of the central propositions of that theory. It was Hilbert who first distilled the norm residue idea from Gauss' theory of genera of quadratic forms and Kummer's highly complicated investigations about the reciprocity law of lth powers (l an odd prime). We begin by developing the relevant facts for quadratic fields.

If a character $\chi(s)$ is defined for the elements s of a group and $\chi(s^2) = 1$ for every element s, then the elements s with the property $\chi(s) = 1$ form either the whole group or a subgroup of index 2. Indeed $\chi(s^2) = 1$ implies

$$\chi^2(s) = 1, \quad \text{i.e.,} \quad \chi(s) = +1 \quad \text{or} \quad -1.$$

Because the quadratic residues modulo an odd prime p form half of the p - 1 residues a prime to p, the division into quadratic residues and non-residues can be described by a character $\left(\frac{a}{p}\right)$. This property of dichotomy is lost if we pass from $\mathfrak{g}_p$ to $\mathfrak{g}(p)$: only 1/4 of the p-adic numbers are squares if p is odd, and only 1/8 if p = 2. The property is restored for norm residues.

Let k be the field of all p-adic numbers, k^* the multiplicative group which results from k by excluding 0, and $\chi(C)$ a character in this group satisfying the relation $\chi(C^2) = 1$ ("quadratic character"). As we are free to multiply C by an even power of p we may suppose C either to be a unit A or of the form pA. Assume p to be odd. Since A is a p-adic square provided

$$A \equiv 1 \pmod{p},$$

$\chi(A)$ depends merely on the residue a of A mod p, and we must have

$$(15.1) \qquad \chi(A) = \chi(a), \qquad \chi(pA) = \delta \cdot \chi(a)$$

with δ independent of a and $\delta = \pm 1$. Of course $\chi(a) = 1$ for all quadratic residues a, and hence $\chi(a)$ is either the principal character χ_0 which = 1 for all the p - 1 elements of the group k_0^* of the residues prime to p, or

$$\chi(a) = \left(\frac{a}{p}\right).$$

If p is 2 this description is to be modified only in that $\chi(A)$ depends on the residue a of A mod 8.

The four odd residues mod 8 form the four-group of which

$$(1, 3), \qquad (1, 5), \qquad (1, 7)$$

each form a subgroup of index 2. Hence besides the principal character χ_0 it possesses three quadratic characters

χ_3, χ_5, χ_7 whose values for the arguments 1, 3, 5, 7 mod 8 are given in the following table:

	1	3	5	7
χ_3	1	1	-1	-1
χ_5	1	-1	1	-1
χ_7	1	-1	-1	1 .

Our χ must be χ_0, χ_3, χ_5, or χ_7.

Let B be a given element of k^* and $K = k(\sqrt{B})$. This K is a field if B is not a p-adic square; but even should this be the case, K always is a commutative algebra (consisting of the k-polynomials of an indeterminate β modulo $\beta^2 - B$). A squared factor in B is of no influence. Hence we may again suppose B to be a unit or p times a unit. An element A of k^* is said to be norm in K if it is norm of an element of K. This obviously means that

$$(15.2) \qquad A = z^2 - By^2$$

has a solution z,y in k. We maintain that there is a quadratic character

$$\chi(A) = \left(\frac{A}{K}\right) = (A,B)$$

in k^* such that $\chi(A) = 1$ if and only if A is norm in $K = k(\sqrt{B})$. As a preliminary to the proof we make the following remarks about the solvability of the equation (15.2) in any field k which is not of characteristic 2.

Let [A,B] indicate the statement that (15.2) has a solution z,y in k. Then

$$[AC^2,B] \sim [A,B], \qquad [A,BC^2] \sim [A,B],$$

C being any element in k^*, i.e., any element $\neq 0$ in k and ~ here denoting logical equivalence. When we make the solution homogeneous by writing $\frac{z}{x}$, $\frac{y}{x}$ for z and y, we are concerned with the solvability of the equation

$$(15.3) \qquad Ax^2 + By^2 = z^2$$

under the restriction $x \neq 0$. The latter can be replaced by the more symmetric restriction

$$(15.4) \qquad (x,y,z) \neq (0,0,0).$$

Indeed if (15.3) has a non-vanishing solution with $x = 0$, one must have $y \neq 0$ and therefore $B = \left(\frac{z}{y}\right)^2$ is a square B_0^2 in k. But then (15.2) or

$$A = (z - B_0 y)(z + B_0 y)$$

is also solvable; take for instance

$$z + B_0 y = A, \qquad z = B_0 y = 1, \qquad \text{or}$$

$$z = \frac{A + 1}{2}, \qquad y = \frac{A - 1}{2B_0} .$$

With the symmetric restriction (15.4) one gets the equivalence

(15.5) $$[A,B] \sim [B,A].$$

Writing (15.3) in the form

$$Ax^2 + By^2 + Cz^2 = 0 \qquad \text{with } C = -1$$

one observes a further symmetry regarding the interchange of A,B and C. However as we do not wish to introduce another argument C, we formulate this symmetry as

(15.6) $$[A,B] \sim [-AB,B]$$

which results from the following form of (15.3):

$$-ABx^2 + Bz^2 = (By)^2.$$

Remember that A and B are both assumed $\neq 0$.

In the following we use this notation: For any finite prime spot p, A,B are p-adic units,

$$K = k(\sqrt{B}), \qquad K' = k(\sqrt{pB}).$$

We set $A \equiv a$, $B \equiv b$ modulo p if p is odd, modulo 8 if $p = 2$. We also include the infinite prime spot $p = \infty$; k is then the field of all real numbers, and A and B are any two real numbers $\neq 0$.

Theorem IV 15, A. *If p is odd then*

$$\left(\frac{A}{K}\right) = 1, \qquad \left(\frac{pA}{K}\right) = \left(\frac{b}{p}\right),$$

or in the notation (15.1):

$$\chi = \chi_0, \; \delta = \left(\frac{b}{p}\right).$$

Furthermore

$$\left(\frac{A}{K'}\right) = \left(\frac{a}{p}\right) = \chi(a), \qquad \left(\frac{pA}{K'}\right) = \delta \cdot \chi(a), \qquad \delta = \left(\frac{-b}{p}\right).$$

If $p = 2$, then for K:

$$\chi(a) = 1 \text{ if } b \equiv 1 \ (4), \; \chi(a) = (-1)^{\frac{a-1}{2}} \text{ if } b \equiv 3 \ (4)$$

and

$$\delta = \left(\frac{b}{2}\right) = (-1)^{\frac{b^2-1}{8}} = +1 \text{ or } -1$$

according as $b \equiv \pm 1$ or $b \equiv \pm 5$ (8); while for K':

$$\chi(a) = 1 \text{ for } a \equiv 1 \text{ or } 1 - 2b \ (8), \; \chi(a) = -1$$

for the other two odd residues mod 8, and

$$\delta = \chi(-b) = \left(\frac{b}{2}\right).$$

For $p = \infty$:

$$\left(\frac{A}{K}\right) = 1 \text{ if } B > 0, \quad \left(\frac{A}{K}\right) = \operatorname{sgn} A \text{ if } B < 0.$$

Proof.

α) p odd prime number.

[α1] $(A,B) = 1.$

One has to show that

(15.7) $$Ax^2 + By^2$$

is capable of representing a quadratic residue; for then the congruence

$$z^2 \equiv Ax^2 + By^2 \quad (p)$$

has a solution $z \not\equiv 0$ (p) and thus the corresponding p-adic equation is solvable. If A is quadratic residue one may simply take $x = 1$, $y = 0$; similarly if B is quadratic residue. If A and B are both quadratic non-residues then (15.7) ranges over the sums of any two non-residues, provided x, y run independently over the residues 1, 2,..., p - 1. If such a sum never were a quadratic residue then it would always be either $\equiv 0$ (p) or a quadratic non-residue. But this is impossible since it would imply that one after the other of the numbers

$$1b, 2b, \ldots, (h+1)b = (hb) + b, \ldots, (p-1)b$$

is quadratic non-residue.

$$[\alpha 2] \qquad (pA,B) = \left(\frac{b}{p}\right).$$

A non-vanishing solution of the homogeneous equation

$$pAx^2 + By^2 = z^2$$

can be assumed to consist of integers x,y,z without a common divisor p. Then y is necessarily not divisible by p. Otherwise $z^2 : p$, hence $z : p$, hence $x^2 : p$, $x : p$. Therefore the equation is possible only if B is quadratic residue. Vice versa, this being the case one obtains a solution with $x = 0$, $y = 1$.

$$[\alpha 3] \qquad (A,pB) = \left(\frac{a}{p}\right)$$

follows from [α2] because of the symmetry (15.5).

$$[\alpha 4] \quad (pA,pB) = \left(\frac{-ab}{p}\right) = \left(\frac{-b}{p}\right) \cdot \left(\frac{a}{p}\right)$$

is a consequence of (15.6):

$$(pA,pB) = (-AB,pB).$$

β) p = 2.

$$[\beta 1] \qquad (A,B) = (-1)^{\frac{a-1}{2} \cdot \frac{b-1}{2}}.$$

If $a \equiv 1$ (4), then

$$C = Ax^2 + By^2$$

becomes $\equiv 1$ (8) for $x = 1$ and $y = 0$ or 2, and then $z^2 = C$ has a 2-adic solution z. Similarly if $b \equiv 1$ (4). If however a and b are both $\equiv 3$ (4) then a solution of

$$Ax^2 + By^2 = z^2$$

in integers x,y,z without a common divisor 2 is impossible because the left side is $\equiv 3$ or 2 (mod 4) for

$$\begin{array}{c|c|c|} x \equiv 1 & 0 & 1 \\ y \equiv 0 & 1 & 1 \end{array} \pmod 2.$$

[β2] $$(2A,B) = 1$$

if either

$$b \equiv 1 \ (8) \quad \text{or} \quad 2a + b \equiv 1 \ (8),$$

else $= -1$. Indeed for the reason explained in the case of odd p we may assume the y of the solution of

$$2Ax^2 + By^2 = z^2$$

to be odd. The possible residues of the left side mod 8 are then b and b + 2a.

[β3] $$(A,2B) = 1$$

if and only if

$$a \equiv 1 \quad \text{or} \quad 1 - 2b \pmod 8.$$

Denoting by $\chi(a)$ this character which is one of the three characters χ_3, χ_5, χ_7 we finally have

[β4]

$$(2A,2B) = (-AB,2B) = \chi(-ab) = \chi(a) \cdot \chi(-b).$$

γ) $p = \infty$. The statements of the theorem about $p = \infty$ are obvious.

After having proved our theorem we specialize B as a rational number b in $\mathcal{g}$ and use the letters k,K to indicate the field $\mathcal{g}$ and the algebra $\mathcal{g}(\sqrt{b})$ respectively. We have introduced a character

$$\left(\frac{A}{K}\right)$$

for the p-adic numbers $A \neq 0$, $\left(\frac{A}{K}\right) = 1$ indicating that A is a norm in K(p). If A is also a p-adic number $I_p(a)$ corresponding to a number a in k we write

$$\left(\frac{I_p(a)}{K}\right) = \left(\frac{a,K}{p}\right) = \left(\frac{a,b}{p}\right)$$

The following universal rules obtain:

$$\left(\frac{a_1a_2,b}{p}\right) = \left(\frac{a_1,b}{p}\right)\left(\frac{a_2,b}{p}\right), \tag{15.8}$$

$$\left(\frac{a,b}{p}\right) = \left(\frac{b,a}{p}\right),$$

and thus also

$$\left(\frac{a,b_1b_2}{p}\right) = \left(\frac{a,b_1}{p}\right)\left(\frac{a,b_2}{p}\right). \tag{15.8'}$$

We now are in a position to verify this important

Theorem IV 15, B.

$$\prod_p \left(\frac{a,b}{p}\right) = 1, \tag{15.9}$$

the product extending to all finite and infinite prime spots p.

Proof. On account of (15.8), (15.8') it suffices to assume a and b to be -1 or a prime number. Hence these possibilities

$$(a,b) = (-1,-1),\ (-1,q),\ (q,q),\ (q,q')$$

where q is a prime number and q,q' are different prime numbers. The case (q,q) is reduced to the preceding one

(-1,q) by the relation

$$(q,q) = (-1,q)$$

which follows from (15.6) and holds with respect to every prime spot p. Distinguishing between odd q and q = 2 we thus have to examine the five cases included in the following table which gives the matrix of the values

$$\left(\frac{a,b}{p}\right)$$

with (a,b) as indicator of the rows and p of the columns:

(a,b) \ p	∞	2	q	q'
(-1,-1)	-1	-1		
(-1, 2)	1	1		
(-1, q)		$(-1)^{\frac{q-1}{2}}$	$\left(\frac{-1}{q}\right)$	
(2, q)		$\left(\frac{q}{2}\right)$	$\left(\frac{2}{q}\right)$	
(q,q')		*	$\left(\frac{q'}{q}\right)$	$\left(\frac{q}{q'}\right)$

$\left(\frac{q}{2}\right)$ is $(-1)^{\frac{q^2-1}{8}}$ and * stands for $(-1)^{\frac{q-1}{2}\cdot\frac{q'-1}{2}}$

q and q' are different (positive) odd prime numbers. The values not given equal 1. The verification of (15.9) is immediate, and the proposition proves equivalent to the quadratic reciprocity law and its supplements.

Here we face a new and important instance of the parallel parts played by the infinite and the finite prime spots. Comparison of formula (15.9) with (5.2) is instructive. We seem to have found a more natural and fundamental form of the reciprocity law.

The question naturally arises under what conditions the rational number a is a norm in $\mathfrak{g}(\sqrt{b})$ or when the equation

$$ax^2 + by^2 - z^2 = 0 \qquad (15.10)$$

has a non-vanishing rational solution (x,y,z). A necessary

condition is certainly that this equation should be locally solvable at every prime spot p, or

$$(15.11) \qquad \left(\frac{a,b}{p}\right) = 1 \quad \text{for every } p.$$

It is true that these conditions together are also sufficient although we shall forego proving it here. This question of the representability of 0 by a given ternary quadratic form (15.10) is of a nature similar to that of the rational transformability of quadratic forms by which in §5 we introduced the infinite prime spot. The fact following from (15.9) that the condition $\left(\frac{a,b}{\infty}\right) = 1$ is redundant besides those referring to the finite spots could easily mislead one to overlook the infinite prime spot. However if one studies the same question in an arbitrary number field instead of $\mathfrak{g}$, one has only one relation (15.9), but in general several infinite prime spots, so that their omission from the set of conditions (15.11) would lead to faulty results.

16. General Norm Residue Symbol and the Theory of Class Fields

Let us try to apply to the general situation the experiences gathered in the quadratic case! We consider an algebraic number field k and a field K over k of relative degree n. For any prime ideal $\mathfrak{p}$ in k we have the field $k(\mathfrak{p})$ of the $\mathfrak{p}$-adic numbers in k and the group $k^*(\mathfrak{p})$ arising from it by exclusion of 0. $\mathfrak{P}$ and $K(\mathfrak{P})$, $K^*(\mathfrak{P})$ have a similar meaning for K. With $\mathfrak{P}$ ranging over the distinct prime divisors of $\mathfrak{p}$ in K, the algebra $K(\mathfrak{p})$ is the direct sum

$$\sum_{\mathfrak{P}|\mathfrak{p}} K(\mathfrak{P}).$$

If an element $A_{\mathfrak{P}}$ of $K(\mathfrak{P})$ is assigned to each of the prime ideals $\mathfrak{P}|\mathfrak{p}$ then the equation (III, 7.5) suggests the following definition of norm:

$$(16.1) \qquad \alpha_{\mathfrak{p}} = \prod_{\mathfrak{P}|\mathfrak{p}} \mathrm{Nm}_{K(\mathfrak{P})/k(\mathfrak{p})} (A_{\mathfrak{P}}).$$

It is clear accordingly what is meant by the element $\alpha_{\mathfrak{p}}$ of $k^*(\mathfrak{p})$ being a norm in K. An nth power $\alpha_{\mathfrak{p}}^n$ is certainly a norm, namely of the assignment:

$$A_{\mathfrak{P}} = \alpha_{\mathfrak{p}} \quad \text{for each } \mathfrak{P} | \mathfrak{p}.$$

We study merely the case of an Abelian field K/k, i.e., we suppose K/k to be Galois with an Abelian group of automorphisms. Kummer's, Hilbert's and all consequent efforts have shown the extreme difficulty of defining directly the norm residue symbols for certain exceptional prime spots $\mathfrak{p}$, in particular for the prime ideals going into the discriminant of K/k. At present the best way seems to be to adopt a direct definition for all prime spots except those of a certain finite set S and then to use the basic law, Theorem IV 15, B, for covering the exceptional spots. We include in S all infinite prime spots and all prime divisors of the discriminant $\mathfrak{d}$ of K/k. Thus we do not bother to extend our definition of norm to the infinite prime spots of k, although this would be very easy indeed.

Because an nth power always is a norm we first study the equation

$$\xi^n = \alpha \tag{16.2}$$

in $k(\mathfrak{p})$. It has a solution if $\alpha \equiv 1$ modulo a sufficient high power of the prime ideal $\mathfrak{p}$. This statement is a special case of what we have proved in Chapter III, §5 about the factorization of polynomials. The exact result is thus:

Lemma IV 16, A. *If n is of order* δ *with respect to* $\mathfrak{p}$, *then (16.2) is certainly solvable provided*

$$\alpha \equiv 1 \quad (\mathfrak{p}^{2\delta+1}).$$

[In particular, if n is not divisible by $\mathfrak{p}$, *the condition* $\alpha \equiv 1$ $(\mathfrak{p})$ *is sufficient.]*

Proof. Let us suppose we have a solution ξ of the congruence

$$\xi^n \equiv \alpha \ (\mathfrak{p}^{l+\delta})$$

with $l > \delta$. We show how to derive from it a solution ξ' of the same congruence with the next higher exponent $l + \delta + 1$, ξ' being congruent to ξ modulo $\mathfrak{p}^l$. Choose a prime number π to $\mathfrak{p}$ and set

$$\xi' = \xi + \pi^l \upsilon.$$

We then find

$$\xi'^n \equiv \xi^n + n\xi^{n-1}\pi^l\upsilon \quad (\mathfrak{p}^{2l}),$$

hence a fortiori

$$\xi'^n \equiv \xi^n + n\pi^l\upsilon \quad (\mathfrak{p}^{l+\delta+1}),$$

and this is $\equiv \alpha(\mathfrak{p}^{l+\delta+1})$ if one chooses

$$\upsilon \equiv \frac{\alpha - \xi^n}{n\pi^l}(\mathfrak{p}).$$

The right side of this congruence is integral at $\mathfrak{p}$.

Induction thus leads to the desired result because the assumption guarantees the solution $\xi = 1$ for $l = \delta + 1$. The solution of (16.2) we have constructed has the property

$$\xi \equiv 1 \quad (\mathfrak{p}^{\delta+1}).$$

For an infinite prime spot where $k(\mathfrak{p})$ is the real or complex field, (16.2) is solvable for any $\alpha \neq 0$ except if $\mathfrak{p}$ is real and n even; then $\alpha > 0$ is the necessary and sufficient condition.

Let now $\mathfrak{p}$ be any finite prime spot which does not divide $\mathfrak{d}$. It decomposes in K into g distinct prime ideals,

$$\mathfrak{p} = \mathfrak{P}_1 \cdots \mathfrak{P}_g$$

of degree f, fg = n. The orders at $\mathfrak{p}$ of each of the partial norms in the right member of (16.1) and hence of the total norm are necessarily multiples of f. Therefore it seems indicated to introduce a primitive f^{th} root of unity ζ and to define for any element $\alpha_{\mathfrak{p}}$ of $k^*(\mathfrak{p})$:

$$\left(\frac{\alpha_{\mathfrak{p}}}{K}\right) = \zeta^i \tag{16.3}$$

where i is the order of $\alpha_{\mathfrak{p}}$ at $\mathfrak{p}$.

However here arises a new difficulty which was absent in the previous section: No way is visible to fix the root ζ in an unambiguous algebraic manner; but as long as we choose freely, for each individual $\mathfrak{p}$, a corresponding primitive f^{th} root of unity there is no chance at all to obtain a universal law of the kind (15.9). Artin overcame this difficulty by introducing what in Chapter III, §10 has been called the Artin symbol: It is an operation $s = \left(\frac{K}{\mathfrak{p}}\right)$ of the Galois group $\mathfrak{G}$ of K/k, uniquely determined by $\mathfrak{p}$, such that the congruence

$$A^s \equiv A^P \qquad (P = N\mathfrak{p})$$

holds for any integer A of K and modulo each of the prime ideals $\mathfrak{P} | \mathfrak{p}$ and hence modulo $\mathfrak{p}$ itself. This element s is of order f. Therefore we replace the tentative definition (16.3) by the final one:

$$(16.4) \qquad \left(\frac{\alpha_{\mathfrak{p}}}{K}\right) = \left(\frac{K}{\mathfrak{p}}\right)^i$$

where i denotes the order of $\alpha_{\mathfrak{p}}$ in $\mathfrak{p}$. The norm residue symbol now maps the group $k^*(\mathfrak{p})$ homomorphically into the Galois group rather than into the group of nth roots of unity. Our definition causes the law $\left(\frac{\alpha_{\mathfrak{p}}}{K}\right) = 1$ to hold for any norm $\alpha_{\mathfrak{p}}$.

Although this is perhaps the more important half of the link between $\mathfrak{p}$-adic norms and the symbol $\left(\frac{\alpha_{\mathfrak{p}}}{K}\right)$ we should feel much more assured of being on the right road if the converse were established also, namely that

$$(16.5) \qquad \left(\frac{\alpha_{\mathfrak{p}}}{K}\right) = 1$$

forces $\alpha_{\mathfrak{p}}$ to be a $\mathfrak{p}$-adic norm. The assumption (16.5) implies that the order of $\alpha_{\mathfrak{p}}$ at $\mathfrak{p}$ is a multiple of f. We know that we can find an integer Π in K whose relative norm $Nm\,\Pi$ is exactly divisible by the fth power of $\mathfrak{p}$. By multiplication with a suitable power of that norm we can reduce $\alpha_{\mathfrak{p}}$ to a unit at $\mathfrak{p}$.

The partial norm on the right side of (16.1) for any $\mathfrak{P}$-adic unit $A_{\mathfrak{P}} = A$ is

$$\prod_{i=0}^{f-1} A^{s^i} \equiv \prod_{i=0}^{f-1} A^{P^i} = A \text{ to the power } \frac{P^f - 1}{P - 1} \pmod{\mathfrak{P}}.$$

Let T be a primitive residue mod $\mathfrak{P}$ so that every $\mathfrak{P}$-adic unit is congruent to a power T^j mod $\mathfrak{P}$. The partial norm in $K(\mathfrak{P})/k(\mathfrak{p})$ of T^j is congruent to

$$T \text{ to the power } j \cdot \frac{P^f - 1}{P - 1} \pmod{\mathfrak{P}}$$

and hence assumes P - 1 values which are incongruent mod $\mathfrak{P}$ and thus mod $\mathfrak{p}$, if j ranges over the exponents 1,..., P - 1. In other words, it takes on <u>all</u> the P - 1 residues in k prime to $\mathfrak{p}$, and each the same number of times, while j varies over its full range

$$j = 1, \ldots, P^f - 1.$$

Let now $A_{\mathfrak{P}}$ for each of the g prime divisors $\mathfrak{P}$ of $\mathfrak{p}$ run over all residues in K prime to $\mathfrak{P}$. The norm $\alpha_{\mathfrak{p}}$ of this assignment as defined by (16.1) then takes on each residue in k prime to $\mathfrak{p}$ the same number of times, namely

$$\frac{(P^f - 1)^g}{P - 1} \text{ times.}$$

What we have found so far is that each $\mathfrak{p}$-adic unit $\beta_{\mathfrak{p}}$ is <u>congruent</u> mod $\mathfrak{p}$ to a norm $\alpha_{\mathfrak{p}}$ in K.

We wish to show that it <u>equals</u> such a norm. To that end we also exclude the prime ideals $\mathfrak{p}$ dividing n. Then the equation

$$\xi^n = \beta_{\mathfrak{p}} / \alpha_{\mathfrak{p}}$$

has a $\mathfrak{p}$-adic solution ξ, and while $\alpha_{\mathfrak{p}}$ is the norm of the assignment $A_{\mathfrak{P}}$, $\beta_{\mathfrak{p}}$ is the norm of the assignment $\xi \cdot A_{\mathfrak{P}}$. We have reached our goal by an argument which is more natural and yields more complete results than the one which we used in the quadratic case for proving [α1] in §15.

At this point it seems advisable to introduce a new concept due to Chevalley for which he uses the word "idèle" as an abbreviation for "ideal element." From the $\mathfrak{p}$-adic standpoint the most essential feature of a number α of k is its associating a $\mathfrak{p}$-adic number $I_{\mathfrak{p}}(\alpha) = \alpha_{\mathfrak{p}}$ with each prime spot $\mathfrak{p}$. Therefore we define an ideal number $\underline{\alpha}$ by associating in arbitrary fashion a $\mathfrak{p}$-adic number $\alpha_{\mathfrak{p}}$ with each prime spot $\mathfrak{p}$. $\alpha_{\mathfrak{p}}$ is called the $\mathfrak{p}$-component of $\underline{\alpha}$. The ideal numbers form a ring in which the field of ordinary numbers is contained. However here we are interested only in their multiplicative, not their additive aspect. The ordinary numbers $\alpha \neq 0$ form a group k*. Accordingly we consider the group J_k of all ideal numbers $\underline{\alpha}$ for which every component $\alpha_{\mathfrak{p}} \neq 0$ and moreover $\alpha_{\mathfrak{p}}$ is a $\mathfrak{p}$-adic unit except for a finite number of prime spots $\mathfrak{p}$; these ideal numbers we call ideal elements or simply <u>elements</u>. k* is contained in J_k as the subgroup of principal elements. Because of the admitted exceptions the definition makes sense even if we do not know what we mean by a $\mathfrak{p}$-adic unit for an infinite prime spot $\mathfrak{p}$. But it is better to come to a definite understanding on this point: For infinite $\mathfrak{p}$ each element of $k^*(\mathfrak{p})$ shall be called a unit. Interpreting $\alpha_{\mathfrak{p}}$ not only as the $\mathfrak{p}$-component of the element $\underline{\alpha}$ but also as the "primary" element whose $\mathfrak{p}$-component is $\alpha_{\mathfrak{p}}$ while every other component equals 1, we may write $\underline{\alpha}$ as the product

$$\underline{\alpha} = \prod_{\mathfrak{p}} \alpha_{\mathfrak{p}} .$$

The advisability of operating with a function $\alpha_{\mathfrak{p}}$ associating a $\mathfrak{p}$-adic number $\alpha_{\mathfrak{p}}$ with each prime spot $\mathfrak{p}$, rather than with the individual components $\alpha_{\mathfrak{p}}$, is pretty evident in our definition (16.1) of norm where several components $A_{\mathfrak{P}}$ appear on the right side, namely all those which correspond to prime divisors $\mathfrak{P}$ of $\mathfrak{p}$ in K. This relation will now simply be written as

$$\underline{\alpha} = \mathrm{Nm}\, \underline{A}$$

($\underline{A}$ an ideal number in K, $\underline{\alpha}$ in k). If $\underline{A}$ is an element of J_K then $\underline{\alpha} = \mathrm{Nm}\, \underline{A}$ is an element of J_k. The necessity becomes even more urgent when we try to formulate the general law corresponding to (15.9). We accept the definition (16.4) for all prime spots $\mathfrak{p}$ outside a certain finite set S which of necessity contains the infinite prime spots and the prime divisors of $\mathfrak{d}$. Let us assume for the moment we have succeeded in extending the definition of $\left(\frac{\alpha_{\mathfrak{p}}}{K}\right)$ to the excluded prime spots $\mathfrak{p}$. For any element $\underline{\alpha}$ of J_k we then introduce

$$\left(\frac{\underline{\alpha}}{K}\right) = \prod_{\mathfrak{p}} \left(\frac{\alpha_{\mathfrak{p}}}{K}\right).$$

The product extending to all prime spots $\mathfrak{p}$ has a meaning because $\left(\frac{\alpha_{\mathfrak{p}}}{K}\right) = 1$ for almost all $\mathfrak{p}$, namely for all $\mathfrak{p}$ outside S for which $\alpha_{\mathfrak{p}}$ is a $\mathfrak{p}$-adic unit. The symbol $\left(\frac{\underline{\alpha}}{K}\right)$ defines a homomorphic mapping of the group J_k into the Galois group $\mathfrak{G}$ of K/k. The law in question states that $\left(\frac{\underline{\alpha}}{K}\right) = 1$ for any principal element $\underline{\alpha}$ of J_k. This is the general reciprocity law in Artin's form. Two elements $\underline{\alpha}$, $\underline{\beta}$ of

J_k are said to be equivalent and will be counted in the same class if $\underline{\alpha}/\underline{\beta}$ is principal. $\left(\frac{\alpha}{K}\right)$ then establishes rather a homomorphic mapping of the group $\mathcal{L}$ of <u>classes</u> in J_k upon $\mathfrak{G}$. This group $\mathcal{L}$ renders a much more complete picture of the arithmetical structure of k than the class group of ideals. Moreover $\left(\frac{\alpha}{K}\right) = 1$ whenever $\underline{\alpha}$ is norm in K. The elements of the form

$$\underline{\alpha} = \gamma \cdot \mathrm{Nm}\,\underline{A}$$

where γ is principal and $\underline{A}$ an element of J_K form a subgroup Nm J_K of J_k. It is to be expected, and this <u>first main theorem</u> is the goal towards which our whole trend of thoughts converges, that <u>the norm symbol</u> $\left(\frac{\alpha}{K}\right)$ <u>establishes a one-to-one isomorphic mapping between the groups</u> $J_k/\mathrm{Nm}\,J_K$ <u>and</u> $\mathfrak{G}$.

However before such results can be obtained one first has to find the general definition of $\left(\frac{\alpha}{K}\right)$. It is based on the following

<u>Lemma IV 16, B</u>. *Given an element $\underline{\alpha}$ of J_k one can ascertain an ordinary number $\rho \neq 0$ of k such that $(\alpha\rho^{-1})_{\mathfrak{y}}$ is an nth power for every $\mathfrak{y}$ in S.*

<u>Proof</u>. Let $\mathfrak{y}$ be a <u>finite</u> prime spot of S, π a prime number to $\mathfrak{y}$, and suppose n is exactly divisible by the power $\mathfrak{y}^{\delta}$ of the prime ideal $\mathfrak{y}$. If $\alpha_{\mathfrak{y}}$ is of order h and

$$\frac{\alpha}{\pi^h} \equiv \alpha_0 + \alpha_1\pi + \dots + \alpha_{2\delta} \cdot \pi^{2\delta} + \dots$$

the beginning of its $\mathfrak{y}$-adic expansion we try to determine ρ such that $\frac{\rho}{\pi^h}$ is an integer at $\mathfrak{y}$ satisfying the congruence

$$\frac{\rho}{\pi^h} \equiv \alpha_0 + \alpha_1\pi + \dots + \alpha_{2\delta} \cdot \pi^{2\delta} \ (\mathfrak{y}^{2\delta+1}). \tag{16.6}$$

Indeed according to Lemma IV 16, A, the quotient $\frac{\alpha_{\mathfrak{y}}}{\rho} \equiv 1 \ (\mathfrak{y}^{2\delta+1})$ would then be an nth $\mathfrak{y}$-adic power. For the finite number of prime ideals $\mathfrak{p} = \mathfrak{p}_1, \ldots, \mathfrak{p}_t$ which are present in S the t simultaneous congruences (16.6) have a common solution ρ of the form

$$\rho = \pi_1^{h_1} \ldots \pi_t^{h_t} \cdot \rho_*$$

with an integral ρ_*. It fulfills all the requirements unless n is even and there are real infinite prime spots $\mathfrak{p}^{(\lambda)}$. In that case the conjugates $\rho^{(\lambda)}$ of ρ have to be positive. ρ_* matters only modulo $\mathfrak{p}_1^{2\delta_1+1} \ldots \mathfrak{p}_t^{2\delta+1}$. Let m be a positive rational integer divisible by this ideal. We may replace ρ_* by $\rho_* + mu$ and then choose the rational integer u so as to ensure positive values for the real conjugates $\rho_*^{(\lambda)} + mu$.

If $\psi(\underline{\alpha}) = \left(\frac{\underline{\alpha}}{K}\right)$ is to depend on the class of $\underline{\alpha}$ only then one must have $\psi(\underline{\alpha}) = \psi(\underline{\alpha}\rho^{-1})$ and thus

(16.7)
$$\psi(\underline{\alpha}) = \prod_{\mathfrak{p}} \left(\frac{(\underline{\alpha}\rho^{-1})_{\mathfrak{p}}}{K}\right)$$

where the prime spots $\mathfrak{p}$ of S are omitted from the product because $(\underline{\alpha}\rho^{-1})_{\mathfrak{p}}$ is an nth power for every $\mathfrak{p}$ in S. If one uses (16.7) as a definition the first thing to be proved is its independence of the choice of the number ρ. This boils down to the following statement which now takes the place of Theorem IV 15, B:

Lemma IV 16, C. *If ρ is a number $\neq 0$ of k which is a $\mathfrak{p}$-adic n^{th} power for the prime spots $\mathfrak{p}$ in S then the product*

$$\prod_{\mathfrak{p}} \left(\frac{\rho_{\mathfrak{p}}}{K}\right) = 1$$

if extended to all prime spots $\mathfrak{p}$ outside S.

Taking this fundamental fact for granted one easily sees that (16.7) has the desired properties:

(i) $\psi(\underline{\alpha\beta}) = \psi(\underline{\alpha}) \cdot \psi(\underline{\beta})$.

(ii) $\psi(\alpha_{\mathfrak{p}}) = \left(\frac{\alpha_{\mathfrak{p}}}{K}\right)$ if $\mathfrak{p}$ is not in S.

(At the left side $\alpha_{\mathfrak{p}}$ stands for the primary element $\alpha_{\mathfrak{p}}$ which =1 for all prime spots $\neq \mathfrak{p}$. For $\underline{\alpha} = \alpha_{\mathfrak{p}}$ the number $\rho = 1$ fulfills the requirement of Lemma C.)

(iii) $\psi(\underline{\alpha}) = 1$ if $\underline{\alpha}$ is principal

(follows by application of Lemma C to the number $\alpha\rho^{-1}$).

(iv) $\psi(\underline{\alpha}) = 1$ if $\underline{\alpha}$ is a norm Nm $\underline{A}$.

(One determines a number $P \neq 0$ in K such that $\underline{A}P^{-1}$ is a $\mathfrak{P}$-adic n^{th} power for all prime spots $\mathfrak{P}$ which lie over the prime spots $\mathfrak{p}$ of S, and then chooses $\rho = \text{Nm } P$.)

This shows how one can circumvent a given finite set S of prime spots. We shall not prove the central lemma except in a special, though decisive case, namely if K is a subfield of a cyclotomic field $k(\zeta)$ over k. Here ζ denotes a primitive m^{th} root of unity, m being any natural number which is prime to the discriminant of k. The field $k(\zeta)$ then is of degree $\varphi(m)$ over k. Let S consist of the infinite prime spots and the prime divisors of m in k. (This set will of necessity include the prime divisors of the discriminant of K/k.) The hypothesis implies that the orders of ρ with regard to the prime ideals in S are multiples of n. Hence after multiplying ρ by the nth power of a suitable number in k we may suppose ρ to be integral and prime to m; it will then satisfy a congruence

$$\rho \equiv \beta^n \ (m), \tag{16.8}$$

and if n be even the real conjugates $\rho^{(\lambda)}$ of ρ will be positive. The lemma is equivalent to the statement that the substitution

$$s' = \prod_{\mathfrak{p} \text{ not in S}} \left(\frac{\rho_{\mathfrak{p}}}{k(\zeta)}\right)$$

of the Galois group $\mathfrak{G}_m$ of $k(\zeta)/k$ lies in the subgroup $\mathfrak{G}'$ to which K belongs. With

$$(\rho) = \mathfrak{r} = \mathfrak{q}_1^{e_1} \mathfrak{q}_2^{e_2} \cdots$$

that product s'

$$= \left(\frac{k(\zeta)}{\mathfrak{q}_1}\right)^{e_1} \left(\frac{k(\zeta)}{\mathfrak{q}_2}\right)^{e_2} \cdots .$$

$\left(\frac{k(\zeta)}{\mathfrak{q}}\right)$ is the substitution changing ζ into $\zeta^{N\mathfrak{q}}$. Hence s' is the substitution

$$\zeta \longrightarrow \zeta^{N\mathfrak{r}}.$$

Now (16.8) implies

$$Nm\ \rho \equiv (Nm\ \beta)^n \pmod{m}$$

If n is even the real conjugates $\rho^{(\lambda)}$ of ρ are positive, hence Nm ρ = $N\mathfrak{r}$ positive; if n is odd it may happen that $N\mathfrak{r}$ = -Nm ρ. But in any case we can write

$$N\mathfrak{r} \equiv (\pm Nm\ \beta)^n \pmod{m}$$

where the sign $\pm$ is irrelevant for even n and has to be properly chosen for odd n. With the rational integer

$$b = \pm Nm\ \beta$$

we form the substitution

$$\zeta \rightarrow \zeta^b$$

and then get $s' = s^n$. Since $\mathfrak{J}_m/\mathfrak{J}'$ is of degree n, the nth power of any element s of $\mathfrak{J}_m$ lies in $\mathfrak{J}'$. Hence s' lies in $\mathfrak{J}'$ as maintained.

This argument covers the quadratic case treated in the last section.

The general Lemma IV 16, C can be established by an ingenious auxiliary construction, originally due to Tschebotaröff and adapted to the present purpose by Artin and Chevalley, which may be described as a crossing of the arbitrary Abelian field K with appropriate cyclotomic fields. In the course of the full proof of the first main theorem one also finds, provided K/k is cyclic, that a number of k is a norm in K if this is locally true everywhere.

So far the whole affair concerns a single given Abelian field K. There is a second part to the theory of class fields which ties the structure of the class group $\mathcal{L}$ of J_k to all possible Abelian fields K over k. Let J_k^* be any subgroup of J_k which shares the following properties with the subgroups Nm J_K corresponding to Abelian fields K over k:

(i) it contains all principal elements;

(ii) there is a natural number n such that all n^{th} powers of elements lie in J_k^*;

(iii) there is a finite set S of prime spots such that a unit lies in J_k^* whenever it equals 1 at the prime spots of S. (The element $\underline{\alpha}$ of J_k is called a unit if it is a unit at every prime spot.)

Such an "admissible" subgroup J_k^* is necessarily of finite index under J_k. Indeed each element $\underline{\alpha}$ of J_k has as its content an ideal

$$\mathfrak{a} = \Pi \mathfrak{p}^e$$

where e denotes the order of $\alpha_{\mathfrak{p}}$ with respect to $\mathfrak{p}$ ($\mathfrak{p}$ any prime ideal). The content of $\rho\underline{\alpha}$ is $\rho\mathfrak{a}$ (ρ any number $\neq 0$ of k). If the contents of $\underline{\alpha}$ and $\underline{\alpha}'$ are equivalent ideals $\mathfrak{a}$ and $\mathfrak{a}'$ then $\underline{\alpha}/\underline{\alpha}'$ is equivalent to a unit. Thus considering the fact that <u>the number of classes of ideals is finite</u> we have merely to prove that the subgroup of those units in J_k which are n^{th} powers at the prime spots of S is of finite index within the group U_k of all units. This follows readily from the corresponding fact for the $\mathfrak{p}$-adic units at an individual prime spot $\mathfrak{p}$.

<u>Second main theorem</u>. *Any admissible subgroup J_k^* of J_k uniquely determines an Abelian field K over k such that $J_k^* = Nm\ J_K$.*

The proof of this theorem must depend on the actual construction of Abelian fields. Abelian fields are composed of cyclic fields. Assuming k to contain a primitive n^{th} root of unity, every cyclic field of degree n over k is generated by the nth root $\sqrt[n]{\gamma}$ of a number γ in k. This is a classical result due to Lagrange; but as Kummer was the first to explore the arithmetic of these fields $k(\sqrt[n]{\gamma})$, they are now called Kummer fields. The simple construction of adjoining an nth root $\sqrt[n]{\gamma}$ puts the Abelian fields at our finger tips.

Chevalley has given a succinct formulation for the whole theory by making use of characters and topology. Let $\chi(s)$ be a character of the Galois group $\mathfrak{G}$ of K/k; its order $\bar{n}$ will be a divisor of the degree n of K/k, and the subgroup of those s for which $\chi(s) = 1$ will determine a subfield Z_χ of K which is cyclic of degree $\bar{n}$ over k. We set at the same time

$$\varphi(\underline{\alpha}) = \chi(s) \quad \text{with} \quad s = \left(\frac{\underline{\alpha}}{K}\right).$$

$\varphi(\underline{\alpha})$ is a character of the group J_k. Thus Chevalley in a certain sense undoes what Artin had accomplished: Instead of being a substitution s, the norm residue symbol $\varphi(\underline{\alpha})$ has again become a complex number ζ on the unit circle (satisfying the equation $\zeta^n = 1$). As one readily verifies, $\varphi(\underline{\alpha})$ stays unaltered when K is replaced by any Abelian field over k that includes Z_χ. This is the reason why Chevalley, not afraid of resorting to arguments of a thoroughly infinitistic nature, introduces the infinite field W/k embracing all finite Abelian fields over k, and its group $\mathfrak{G}_W$ which is commutative. Any subgroup $\mathfrak{G}'$ of $\mathfrak{G}_W$ of finite index is considered a neighborhood of 1 in $\mathfrak{G}_W$. A continuous character $\chi(s)$ of $\mathfrak{G}_W$ in the sense of this topology is necessarily of finite order. Indeed any multiplicative group of numbers χ on the unit circle, if it does not consist of $\chi = 1$ alone, contains numbers χ such that

$$|\chi - 1| \geqq \sqrt{3}.$$

Hence if $\mathfrak{G}'$ is a neighborhood of 1 in $\mathfrak{G}_W$ such that for all elements s in $\mathfrak{G}'$ the inequality

$$|\chi(s) - 1| < \sqrt{3}$$

holds, then $\chi(s) = 1$ identically in $\mathfrak{G}'$. The salient point in this argument is that the neighborhoods $\mathfrak{G}'$ are groups. Hence χ is a character of the finite group $\mathfrak{G}_W/\mathfrak{G}'$.

For similar reasons one has to endow the group J_k with this topology: Let S be any finite set of prime spots and n any natural number, U^S the group of units in J_k which = 1 at the prime spots of S, and J^n the group of all n^{th} powers in J_k; then

$$U^S J^n = G^{(S,n)}$$

is a neighborhood of 1. Again the neighborhoods are groups. Notice that

$$G^{(S \cup S', nn')} \subset (G^{(S,n)} \cap G^{(S',n')}).$$

A continuous character $\varphi(\underline{\alpha})$ of J_k in the sense of this topology is necessarily of finite order because for all elements $\underline{\alpha}$ in a certain neighborhood of 1 one will have $\varphi(\underline{\alpha}) = 1$, hence a fortiori

$$(\varphi(\underline{\alpha}))^n = \varphi(\underline{\alpha}^n) = 1$$

for every element $\underline{\alpha}$ of J_k whatsoever. Characters $\varphi(\underline{\alpha})$ which satisfy the equation $\varphi(\underline{\alpha}) = 1$ for all principal elements $\underline{\alpha}$ are called differentials by Chevalley. In this terminology the essential facts of the theory of class fields may be summarized in the one statement that the norm residue symbol establishes a one-to-one isomorphic correspondence between the continuous differentials of J_k on the one side and the continuous characters of $\mathfrak{G}_W$ on the other side. In this way the class group of J_k which is an intrinsic property of k reflects the structure of the edifice of all Abelian fields which can be erected over k.

Since Hilbert first blazed the trail, enormous progress has been made in the theory of class fields, by Furtwängler, Takagi, Hasse, Tschebotaröff, Artin, Chevalley (to mention but the most important names). In particular the transcendental methods of which Hilbert made ample use have been pushed back step by step until they were entirely eliminated. But in spite of all efforts I have the impression that the theory has not yet assumed its final form.

ERRATA

Page 25, line 18. For "factor $\mathfrak{y}/\mathfrak{y}'$," read "factor group $\mathfrak{y}/\mathfrak{y}'$."

Page 40, line 1. For "exist," read "exists."

Page 158, line 17 from bottom. For "finite," read "infinite."

AMENDMENTS

1) [page 66] §11 of Chap. II serves no other purpose than to make clear the relationship of Dedekind's ideals to Kronecker's divisors, from our viewpoint of a theory of divisibility. But even so, I should have mentioned v. d. Waerden's and Artin's theory of quasi-equality of ideals (v. d. Waerden, Moderne Algebra II, Berlin 1931, §103) which throws much light on our discussion of the fields of algebraic functions of several variables. In fairness to Dedekind I should also have pointed out that the three preceding sections in v. d. Waerden's book develop the classical Dedekind theory of ideals without resorting to indeterminates; the procedure is based on E. Noether's and W. Krull's work.

2) [page 168] Chap. IV, §8: In construction of relative units I follow in the main Minkowski and Chevalley. Chevalley's "La theorie de corps de classes," Annals of Mathematics, 41 (1940) pp. 394-418, covers the same material as his course mentioned in the preface. An insight into the structure of the group of relative units is an essential prerequisite for that theory (cf. §16).